THE COMMONWEALTH AND INTERNATIONAL LIBRARY
Joint Chairmen of the Honorary Editorial Advisory Board
SIR ROBERT ROBINSON, O.M., F.R.S., LONDON
DEAN ATHELSTAN SPILHAUS, MINNESOTA

SELECTED READINGS IN PHYSICS
General Editor: D. TER HAAR

SPECIAL THEORY
OF RELATIVITY

SPECIAL THEORY OF RELATIVITY

by
C. W. KILMISTER
Professor of Mathematics, King's College, London

PERGAMON PRESS
Oxford · New York
Toronto · Sydney · Braunschweig

PERGAMON PRESS LTD.,
Headington Hill Hall, Oxford

PERGAMON PRESS INC.,
Maxwell House, Fairview Park, Elmsford, New York 10523

PERGAMON OF CANADA LTD.,
207 Queen's Quay West, Toronto 1

PERGAMON PRESS (AUST.) PTY. LTD.,
19a Boundary Street, Rushcutters Bay, N.S.W. 2011, Australia

VIEWEG & SOHN GMBH,
Burgplatz 1, Braunschweig

Copyright © 1970 Pergamon Press Ltd.

All Rights Reserved. No part of this publication may be reproduced, stored in a retrieval system, or transmitted, in any form or by any means, electronic, mechanical, photocopying, recording or otherwise, without the prior permission of Pergamon Press Ltd.

First edition 1970
Library of Congress Catalog Card No. 78–104961

Printed in Hungary

This book is sold subject to the condition
that it shall not, by way of trade, be lent,
resold, hired out, or otherwise disposed
of without the publisher's consent,
in any form of binding or cover
other than that in which
it is published.

08 006995 9 (flexicover)
08 006996 7 (hard cover

Contents

INTRODUCTION vii

ACKNOWLEDGEMENTS ix

Part I

Chapter 1 Introduction: 1632–1905 3
Chapter 2 Einstein's Contribution 14
Chapter 3 Elementary Consequences of the Lorentz Transformation 34
Chapter 4 Applications in Quantum Theory 57

REFERENCES 87

Part II

1. The Relative Motion of the Earth and the Luminiferous Ether *by* ALBERT A. MICHELSON 91
2. On Kinematic and Mechanical Modes of Representation of the Activity of the Aether from *Aether and Matter*, *by* J. LARMOR 105
3. Electromagnetic Phenomena in a System Moving with any Velocity less than that of Light *by* H. A. LORENTZ 119
4. The Dynamics of the Electron *by* H. POINCARÉ . . . 145
5. On the Electrodynamics of Moving Bodies *by* A. EINSTEIN 187
6. On the Electric Effect of Rotating a Magnetic Insulator in a Magnetic Field *by* MARJORIE WILSON and H. A. WILSON 219

7. Fresnel's Coefficient for Light of Different Colours *by* PROFESSOR P. ZEEMAN 229
8. The Quantum Theory of the Electron *by* P. A. M. DIRAC 237
9. Energies of Cosmic-ray Particles *by* CARL D. ANDERSON 257
10. On Unitary Representations of the Inhomogeneous Lorentz Group *by* E. WIGNER 267

INDEX 297

Introduction

SPECIAL relativity is not, like other scientific theories, a statement about the matter that forms the physical world, but has the form of a *condition* that the explicit physical theories must satisfy. It is thus a *form* of description, playing to some extent the rôle of the grammar of physics, prescribing which combinations of theoretical statements are admissible as descriptions of the physical world. So, to describe it, one needs also to describe those specific theories and to say how much they are limited by it.

But not all physical theories are on the same footing in this respect; for special relativity fulfils the purpose just stated by demanding that our measurements of space and time must be given an operational definition. As far as measuring points of space at one instant of time is concerned, this is no great difficulty; and the idea of time-ordering of events at one place (no matter how complicated this may turn out to be) is taken by special relativity as a given concept, not to be further analysed. But the relation of this time-ordering with that of events at different places is taken to be the central difficulty which has to be explained by operational means. It does this by defining the values of the time variable at distant events in terms of the times of emission and reception of light signals. As a consequence the theory of electromagnetism, which is known to describe the transmission of light very accurately, plays a special role in the theory. Accordingly only the briefest introduction to this has been given and it is assumed that the reader has some familiarity with Maxwell's equations.

The other physical theories which have been most relevant for special relativity have been mechanics, both classical and quantum. Here a good deal has been provided in the way of an introduction, partly because the technique of applying special relativity to these subjects requires their development in directions strikingly different from their usual applications. To some extent this has resulted in an overlapping of this book with others in the series (ter Haar, 1967; Ludwig, 1968) but it is felt that such an overlap is preferable to leaving inexplicable gaps in the exposition of what is actually a tightly knit logical theory.

It is natural that a theory that has as its object the limitation of other methods of description of the physical world will be ubiquitous. It has only been possible to include a few illustrative examples of the applications of special relativity in various fields, and the most striking ones have been selected. The reader who is new to the subject may note that the concluding sections of each of the last three chapters are of considerably greater difficulty than the major part of these chapters, but they are not needed for a study of the major part of the later chapters, so that a preliminary study of the book can consist of Chapter 1, the first three-quarters of each of Chapters 2 and 3, and a brief reading of parts of Chapter 4. The reader may then, if he wishes, return to the omitted portions; those of Chapters 2 and 3 are needed for the mathematical part of Chapter 4.

Acknowledgements

MY THANKS are due to the following bodies for permission to reprint the extracts in the volume:

For Extract 1, the *American Journal of Science*.
For Extract 2, Cambridge University Press.
For Extracts 3 and 7, Koninkligke Nederlandse Akademie van Wetenschappen, Amsterdam.
For Extract 4, Rendiconti del Circolo Matematico di Palermo.
For Extract 5, Johann Ambrosius Barth.
For Extracts 6 and 8, The Royal Society.
For Extract 9, American Physical Society.
For Extract 10, Editors of *Annals of Mathematics*.

PART I

CHAPTER 1

Introduction: 1632—1905

THE special theory of relativity, which is the subject of this book, came into existence in 1905 as a result of the union of two previously unrelated ideas, the notion that motion has a relative character (a phrase which will be explained below), and the notion that optics and mechanics are not two independent disciplines, but must be rendered consistent one with the other.

The fact that motion is not an absolute property of a body but a relation between a body and an observer was known at least as early as Galileo. In his most important work, which he began writing in 1626 (Galileo, 1632) (though he had discussed the possible contents of the book with members of the Church, including a future Pope, as early as 1624), he asks the reader to imagine the following thought experiment. The observers are shut up in the main cabin below decks on a large ship equipped with such things as flying insects, and a bowl of water with fish in. The behaviour of all these things is noted when the ship is at rest and again when the ship is proceeding in a uniform way. It is observed that in uniform motion everything behaves just as at rest. The flying insects do not have to work harder when flying from the stern to the bows; the water does not spill from the bowl, and in general every feature of the phenomena at rest is reproduced in a state of uniform motion. In this experiment Galileo contradicts, and intends to contradict, the views on mechanics of Aristotle, who considered that uniform motion was a quite different state from rest, needing some continual intervention to produce it.

4 SPECIAL RELATIVITY

These considerations of Galileo were taken account of by Newton (1686) who in his *Principia* adheres to the idea of absolute space and absolute time but introduces also the concepts of relative space and relative time. This is to say, he supposes that it is possible to describe motion in terms of a change of absolute space in an interval of absolute time, although what we measure are only the corresponding relative quantities. This section of the *Principia* ends: "But how we are to obtain the true motions from their causes, effects, and apparent differences, and the converse, shall be explained more at large in the following treatise. *For to this end it was that I composed it.*"*

We can make little progress further than this without considering the interaction between mechanics and optics. The earliest detailed discussion of this relation was given by Euler (1750) in a paper which he wrote in 1739. At that time there was a controversy between the wave and particle theories of light. The question was whether light consisted of a beam of small particles or whether it consisted of wave motion in some medium which did not have other physical properties by which it could be detected. The particular result in which Euler noticed a critical difference between the theories was the measurement of aberration by Bradley (1728).

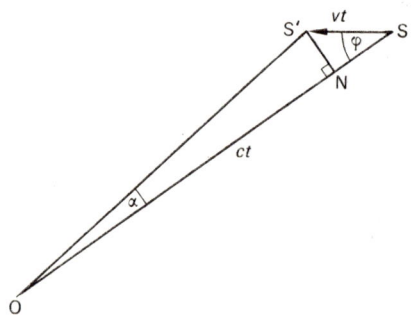

Fig. 1

* The italics are mine. I am indebted to Mr. A. Orr for drawing my attention to this important sentence. C. W. K.

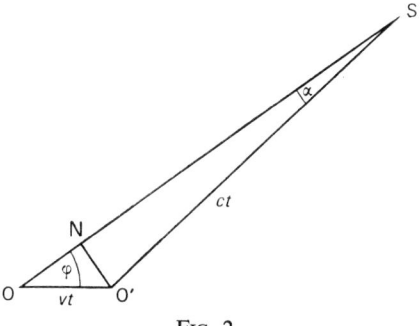

Fig. 2

Aberration is the name given to the change in the measured position of the stars because of the motion of the Earth. Euler first considers the problem of the reception of light from a moving source. If the source has a speed v, and c is the speed of light, Fig. 1 shows that the beam of light emitted at time $t = 0$ from the source S is received at the observer O at a time at which the source will have moved to a new point S'. From the figure it is then clear that the angle of aberration α is given by

$$\tan \alpha = \frac{S'N}{ON} = \frac{v \sin \phi}{c - v \cos \phi}. \tag{1}$$

This result, however, is the one for a moving source and an observer at rest, whereas Euler requires the one in which the source is at rest and the observer moves. He first derives this by making a transformation, i.e. applying to source and observer the same speed $-v$. The source is now reduced to rest and the observer moves; the same angle α arises as before.

Euler then goes on to "investigate the effect which a light ray exercises on the eye in motion", in other words, to carry out a new and independent investigation of aberration by combining the speed of light as a wave motion in the medium with the speed of the observer. The corresponding figure here is Fig. 2, in which O' is the position to which the observer has moved in the time which elapses between the transmission of the signal from S and its reception. The angle α is now given by

$$\tan \alpha = \frac{O'N}{NS} = \frac{v \sin \phi}{\sqrt{(c^2 - v^2 \sin^2 \phi)}}. \tag{2}$$

This result differs from the one found earlier by Euler. He therefore goes on to consider why different results arise here, and he observes at once that in the first case the velocity of light has been assumed to be the same relative to the observer in each case, whereas, in the second, the velocity of light relative to the observer is changed by the motion of the observer in the medium. Accordingly the medium needs to be included and so *if the motion of light is a wave motion in a medium* the two answers (source moving or observer moving) are both right, but in the ballistic theory, in which the light is regarded as a stream of particles, no medium is needed, and so the first derivation is the only correct one. Euler therefore noticed the possibility of an experimental check on the question of whether a ballistic or a wave theory of light was correct, and so of a question concerning the relationship between mechanics and optics.

Euler was only the first (though by far the earliest) of a long series of investigators of the borderland between mechanics and optics. We shall consider here in detail an experiment first performed by Fizeau (1859) to test a theory of Fresnel. It is not easy to attain high accuracy with this experiment; a better accuracy comes with the experiment of Michelson and Morley, discussed below.

In 1810 Arago studied the refraction of light through a prism when the light had come from different stars and was then received in a telescope. If the light consists of a wave motion in a medium through which the Earth is moving, the amount of refraction in the prism ought to depend on the direction in which the telescope is pointed, because the light from different directions will have different speeds relative to the telescope. In fact no such difference is found, and Arago wrote to Fresnel about this. Fresnel replied by drawing heavily on the analogy of acoustics and deducing from this analogy the way in which a solid material could be expected to drag along a sound wave. In Fizeau's experiment the speed of light is measured when it passes through a long pipe containing water. The value of the speed is compared in the two cases when the water is at rest, and when it is moving with speed v. If c is the speed of light in air as before, the speed when the water is at rest is given by c/n where n is the refractive index of water (about $\frac{4}{3}$). The formula which Fresnel predicted by analogy with acoustics, and which is consistent with the results of Arago's experiment, is

$$\text{Velocity} = \frac{c}{n} + v\left(1 - \frac{1}{n^2}\right), \tag{3}$$

showing that the light is dragged on by the water at a slower velocity than the water has itself.

The experiment which was the most subtle attempt to detect the motion of the Earth with respect to the medium of transmission of light (and so with respect to the preferred inertial frame at rest) was that of Michelson (1881) (Extract 1 of the present book), which was repeated later with technical improvements by Michelson and Morley (1887). The experiment was repeated by a number of subsequent workers with complete agreement, until Miller (1925) found considerable variations. However, subsequent performances of the experiment lead to the fairly firm conclusion that Miller's results are due to some unknown experimental error. (See the discussion of Miller's results in R. S. Shankland *et al.* (1955).)

Shorn of experimental difficulties, Michelson's experiment involves splitting a beam of light and transmitting the two halves along two directions at right angles. These beams are reflected at the end of two fixed arms of equal length and return to their starting-point. Any *change* in the difference of times (if any) along the two arms will then cause a shift in the interference fringes. That

8 SPECIAL RELATIVITY

the existence of a medium of transmission through which the apparatus moves does involve such a difference can be seen by considering the special case in which the velocity of the apparatus is along one arm (Fig. 3). Here the dotted lines show the paths of the light, and three successive positions of the apparatus are shown.

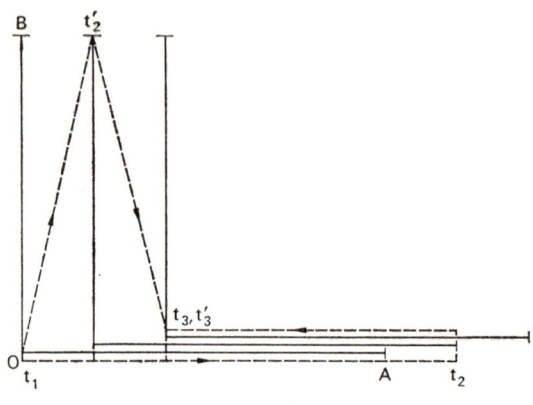

Fig. 3

For the arm OA, if t_1, t_2, t_3 are times of emission, reflection and reception, then
$$c(t_2-t_1) = l+v(t_2-t_1)$$
$$c(t_3-t_2) = l-v(t_3-t_2)$$
so that
$$t_3-t_1 = \frac{l}{c-v}+\frac{l}{c+v} \simeq \frac{2l}{c}\left(1+\frac{v^2}{c^2}\right).$$

On the other hand, for the arm OB, if t_2' is the instant of reflection,
$$c^2(t_2'-t_1)^2 = l^2+v^2(t_2'-t_1)^2,$$
so
$$t_3-t_1 = 2(t_2'-t_1) = \frac{2l}{(c^2-v^2)^{1/2}} \simeq \frac{2l}{c}\left(1+\frac{1}{2}\frac{v^2}{c^2}\right).$$

In Michelson's experiment the apparatus is set up and watch kept for fringe shifts when it is rotated through 90°. None were observed; and since this might be due to the Earth being stationary at that time, the experiment was repeated after 6 months, again with a null result.

Even before the astonishing and paradoxical result of Michelson, the wave theory of light (some form of which seemed absolutely essential to account for interference and diffraction, to say nothing of polarisation) had taken a definitive form, which served to pin-point the difficulties very clearly. Riemann had noticed that the expression $1/\sqrt{(\mu\varkappa)}$, which occurs in transforming from one system of electrical units to another ($\varkappa$ = dielectric constant, μ = magnetic permeability), has for vacuum a value near to the velocity of light (and has the dimensions of velocity). That this was no coincidence was strongly emphasised by the measurements by Weber and Kohlrausch (1856) and Maxwell was greatly influenced by Riemann's coincidence. Unfortunately, Riemann had no theoretical means of establishing a wave theory, and could only suggest generalising Laplace's equation

$$\nabla^2\phi = 0$$

to
$$\nabla^2\phi - \mu\varkappa\frac{\partial^2\phi}{\partial t^2} = 0,$$

a generalisation which satisfactorily gives a velocity to the waves and explains the existence of interference, but is quite inadequate to discuss polarisation.

Maxwell's *Treatise* appeared in 1873, though he had been working on the ideas for much longer (Maxwell, 1873). We sketch a modified form of his arguments briefly. We may begin by noting that the field is described by four vectors: the electric field **E**, the electric induction **D**, the magnetic field **H** and the magnetic induction **B**. The e.m.f. in a closed circuit is proportional to the rate of change of the flux of induction through the circuit:

$$\oint \mathbf{E}.d\mathbf{r} = -\frac{\partial}{\partial t}\int \mathbf{B}.d\mathbf{S},$$

the negative sign resulting from an application of Lenz's law (Lenz, 1834). This equation is a sensible one, since the value of the right-hand side is independent of the particular surface filling the circuit, i.e.

$$\oint \mathbf{B}.d\mathbf{S} = 0$$

for a closed surface (a statement equivalent to saying that there are no free magnetic poles). The differential form of these equations is then easily seen to be

$$\text{curl } \mathbf{E} = -\frac{\partial \mathbf{B}}{\partial t}, \quad \text{div } \mathbf{B} = 0. \tag{4}$$

It is natural to try to have a corresponding expression for

$$\oint \mathbf{H}.d\mathbf{r}, \quad \text{or equivalently} \quad \text{curl } \mathbf{H}.$$

Unfortunately **D**, by Gauss' theorem in electrostatics, has the property

$$\int \mathbf{D}.d\mathbf{S} = 4\pi \int \varrho \, d\tau,$$

i.e.
$$\text{div } \mathbf{D} = 4\pi\varrho. \tag{5}$$

However, this provides exactly the opportunity needed to include also the magnetic effect of currents, since, if we write

$$\text{curl } \mathbf{H} = \frac{\partial \mathbf{D}}{\partial t} + 4\pi \mathbf{J}, \tag{6}$$

we have
$$4\pi \left(\text{div } \mathbf{J} + \frac{\partial \varrho}{\partial t} \right) = 0,$$

which is simply the conservation equation for currents. Equations (4), (5), (6) are Maxwell's equations in their differential form.

In free space, where there are no currents, the so-called constitutive relations

$$\mathbf{B} = \mu \mathbf{H}, \quad \mathbf{D} = \varkappa \mathbf{E}$$

give

$$\operatorname{curl} \operatorname{curl} \mathbf{H} = \varkappa \frac{\partial}{\partial t} \operatorname{curl} \mathbf{E}$$

$$= -\mu\varkappa \frac{\partial^2 \mathbf{H}}{\partial t^2},$$

but $\operatorname{curl} \operatorname{curl} \mathbf{H} = \operatorname{grad} \operatorname{div} H - \nabla^2 H$,

so that
$$\mu\varkappa \frac{\partial^2 \mathbf{H}}{\partial t^2} - \nabla^2 \mathbf{H} = 0,$$

with a similar equation for E, giving a vector form of the scalar wave equation guessed by Riemann. The fact that the equation arises automatically from the basic equations is a great triumph for Maxwell's theory; moreover, the occurrence of the vectors $\mathbf{E}$, $\mathbf{H}$ rather than a scalar ϕ turns out to be sufficient to account for the observed phenomena of polarisation.

But the clarity of Maxwell's theory highlights two great difficulties. Let us first consider plane waves advancing along the x-axis, so that $\mathbf{E}$, $\mathbf{H}$ are both functions of $t - x/c$ where $c = (\mu\varkappa)^{-1/2}$. It follows that

$$\frac{1}{c} \frac{\partial \mathbf{E}}{\partial t} = -\frac{\partial \mathbf{E}}{\partial x}, \quad \text{and} \quad \frac{\partial \mathbf{E}}{\partial y} = \frac{\partial \mathbf{E}}{\partial z} = 0,$$

with similar equations for $\mathbf{H}$. The equations (4), (6) give

$$\frac{\partial H_1}{\partial t} = 0, \quad \frac{\partial E_1}{\partial t} = 0,$$

so that E_1, H_1 are only constant fields superimposed on the wavefield, and so may be set equal to zero. The remaining, varying, field has therefore only y and z components, i.e. it is *transverse* to the direction of propagation. This was a severe obstacle to the builders of mechanical models of the electrical phenomena, for these models had been based on a supposed analogy with sound, in which the vibrations are longitudinal, not transverse. A medium to support wholly transverse vibrations has to have elastic properties of a very odd kind—indeed to have almost the opposite properties to any known substance. None the less, more subtle models were made—Larmor (1900) succeeded in devising a model made by an immense number of small gyroscopes (see Extract 2 of the present book). But Kelvin himself noticed that such model making led nowhere—it never gave rise to any physical hypothesis (see Hesse (1955)). And so the easiest way to avoid the difficulties in it is not to do it.

The other difficulty is much more serious. It is merely the theoretical aspect of the experimental result afterwards found by Michelson; the equations appear to predict a velocity $1/\sqrt{(\mu\varkappa)}$ for the waves independent of the velocity of the coordinate system. And this is paradoxical since one expects to have the Newtonian law of transformation

$$\mathbf{r} \to \mathbf{r}' = \mathbf{r} - \mathbf{v}t$$

giving, for velocities,

$$\mathbf{V} \to \mathbf{V}' = \mathbf{V} - \mathbf{v}.$$

Without going into more details the great difficulty is at once apparent. The speed predicted for light according to this theory is determined only by the ratio of the units employed, and so is independent both of the source and of the observer. It seems as if Maxwell's equations will only apply in a frame of reference which is at absolute rest in the sense in which the term is used by Newton. Absolute rest will now mean at rest relative to the medium at which light is transmitted. But this result, which would have been congenial to Newton as determining his absolute space, had ceased to be at all consistent with mechanics by the nineteenth century. To see why this is so we must return for a minute to the discussion given by Galileo above. The emphasis we want to put here is on the necessity for the motion of the ship to be *uniform*. Subject to this limitation, the thought experiment shows that no absolute space can ever be determined by Newtonian mechanics. All that mechanics can determine would be a collection of reference frames moving uniformly relative to each other, and for all of which the laws of mechanics would be equally valid. Such reference frames are called inertial frames, and everything in mechanics serves to confirm the complete equivalence of all inertial frames.

Although the idea of an inertial frame is not explicit in the usual derivation of Maxwell's equations it is implicit, and so the result of the electromagnetic theory of light seems to be that one particular inertial frame is marked out as the one in which Maxwell's equations apply. Such a situation is an extremely puzzling one and the obvious conclusion is that some changes need to be made so that Maxwell's equations apply in all inertial frames in the same way as Newton's.

CHAPTER 2

Einstein's Contribution

THE unique contribution of Einstein (1905) (reprinted as Extract 5 of the present book) in the discussion of the relationship between mechanics and optics was in directing attention to the need for a proper operational definition of simultaneity for distant events. Instead of dealing with the details of Maxwell's equations at the beginning, he starts from an entirely new physical principle, and reaches a position corresponding to the conventional standpoint later on. Let us consider, he says, the problem of assigning a time to an event distant from the observer. In order to determine the position of the event relative to the order of events happening at the observer it is necessary to relate the distant event in some way to these nearby events, and according to Einstein this can only be done by signalling. Different kinds of signals can be employed but we have reason to expect that the one particular kind of signal, which will be more appropriate than the others, is a light signal, since its transmission is governed by Maxwell's equations, which predict a velocity for it independent of the source and of the observer. Until we have given some rule for assigning time to distant events the concept has no meaning, and so it is necessary to formulate a *rule* for the time assigned to a distant event in terms of the time of transmission of a light signal which will just reach the distant event and the time when it returns.

Einstein defines the time to be assigned to the distant event as the average of these two times. It is necessary to consider a little further the status of this assumption. Certainly, since the concept of time for the distant event has no meaning until we define it, Einstein is at liberty to make this definition if he wishes. On the other hand, we would expect the time allotted to distant events to fulfil certain general conditions, for example, that of being independent of the zero chosen for time reckoning and of the units, so that if the times of emission and reception of the signal are multiplied by a constant, or if a constant is added, the same will be true of the time assigned to the distant event.

In order to assign such a time we may suppose that an observer sends a signal at time t_1 which is reflected at the event and returns to him at time t_3. The convention adopted by Einstein is then that the time assigned to this distant event by the observer should be $\frac{1}{2}(t_1+t_3)$. Now in the first place it is certainly not the case that *any* value could have been assumed for the time, since many of our physical theories would not work unless the time assigned was somewhere between t_1 and t_3. On the other hand, the value to be assigned is not determined by experiment, since we cannot carry out an experiment about the time of distant events until we have formulated a consistent way of assigning such a time. The problem of exactly how much freedom was open to Einstein on this point has been considered very fully by Whitrow (1961). Suppose that the time assigned by the observer to the distant event is $t_2 = f(t_3, t_1)$. If the observer chooses to use a different unit for his time reckoning (say hours instead of seconds), or if he chooses a different zero for the time-reckoning, the same change must result with the assigned time, and we can formulate this condition generally in the form

$$f(kt_3+l, kt_1+l) = kf(t_3, t_1)+l.$$

In order to find what solutions of this equation are possible we may, without loss of generality, make the substitution

$$f(t_3, t_1) = \phi\left(\frac{t_3+t_1}{2}, \frac{t_3-t_1}{2}\right).$$

16 SPECIAL RELATIVITY

Then the condition becomes $\phi(kx+l, ky) = k\phi(x, y)+l$, so that, in particular,

$$\phi(x+l, y) = \phi(x, y)+l,$$

i.e. $\phi(x, y) = x+\psi(y)$, where $\psi(y) = \phi(0, y)$,

and where $\psi(ky) = k\psi(y)$,

so that $\psi(y) = y\psi(1)$.

Hence $\phi(x, y) = x+\varepsilon y$, where ε is constant,

giving $t_2 = f(t_3, t_1) = \frac{1}{2}(t_3+t_1)+\frac{1}{2}\varepsilon(t_3-t_1).$

The previous limitation on the time would now require the constant ε to lie between 1 and -1. However, a further limitation arises if we consider also the assigning of distance to distant events. If the observer estimates the distance of the events to be $g(t_1, t_3)$, and if he supposes that distances along a straight line add up in the usual way so that

$$g(t_1, t_2)+g(t_2, t_3) = g(t_1, t_3),$$

it follows at once that the assigning of distances is done according to the function

$$g(t_1, t_2) = h(t_2)-h(t_1),$$

where $h(t) = -g(t, t_3).$

As a result we have the equation

$$h(t_2) = \tfrac{1}{2}[h(t_1)+h(t_3)]$$

and the important feature of this equation is that the time assigned to the distant event is determined *symmetrically* in terms of the times of transmission and reception of the signal. This therefore restricts the constant ε to a unique value, viz. zero, and we have Einstein's convention. In this case $h(t) = ct$, where c is some constant, which can then be identified with the speed of light.

We can now quickly work out the results of this convention (Milne, 1948). We may suppose that the measured time of the distant event according to an observer who was coincident with the original observer at time $t = 0$, and who is moving with a uniform speed (see Fig. 4), is t_2'. It is clear that the time at which the signal reaches the distant event is proportional to the time at which it was transmitted, so that we have $t_2' = kt_1$. Applying this same argument, however, to the returning signal gives $t_3 = kt_2'$.

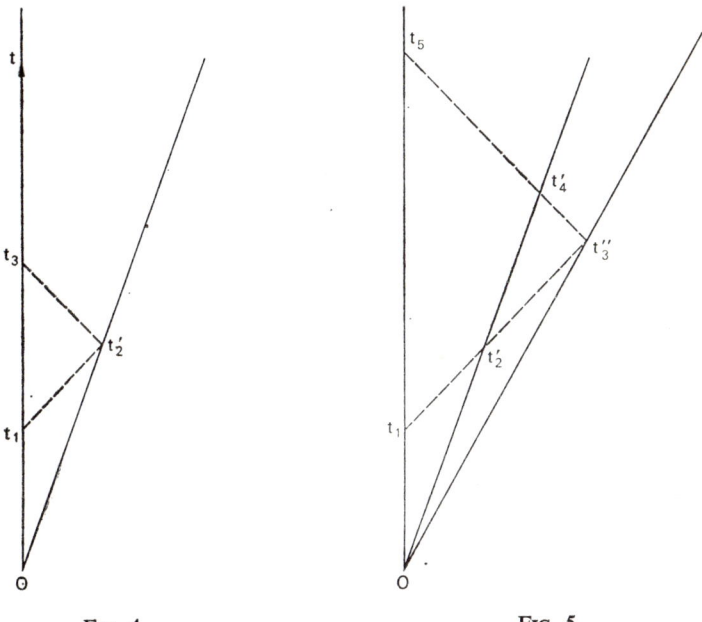

Fig. 4   Fig. 5

As a consequence of these equations it follows that

$$t_3 = k^2 t_1,$$

so that
$$t_2 = \frac{1}{2}(k^2+1)t_1 = \frac{1}{2}\left(k+\frac{1}{k}\right)t_2',$$

$$x = \frac{1}{2}(k^2-1)ct_1, \quad \frac{v}{c} = \frac{x}{ct_2} = \frac{k^2-1}{k^2+1}.$$

Imagine now two observers moving with different speeds relative to the initial observer (see Fig. 5).

From the equations
$$t'_2 = k_1 t_1$$
$$t''_3 = k_2 t'_2,$$
it follows that
$$t''_3 = k_1 k_2 t_1,$$

so that the multiplying constants for the two observers combine according to a simple law, although a different law from the Newtonian combination of velocities. However, using the expression
$$k = \sqrt{\left(\frac{1+v/c}{1-v/c}\right)}$$
it follows at once that the resultant of two velocities will now be expressed by the formula
$$\frac{v_1 + v_2}{1 + v_1 v_2 / c^2}.$$

We may apply this to the experiment of Fizeau. According to an observer at rest with respect to the fluid, the speed of light is c/n. Accordingly for the observer at rest in the laboratory it is
$$\frac{c/n + V}{1 + V/nc} \simeq \left(\frac{c}{n} + V\right)\left(1 - \frac{V}{nc}\right) \simeq \frac{c}{n} + V\left(1 - \frac{1}{n^2}\right),$$

in complete agreement with the observed values.

From the point of view of later developments of the theory it is necessary to rewrite the formulae somewhat, deriving the form in agreement with Einstein's. The transformation between the measurements of one observer and those of another moving uniformly relative to him with speed V can be derived from Fig. 5, and we have
$$t'_2 = k t_1$$
$$t_5 = k t'_4,$$

so that the coordinates (t, x), (t', x') assigned to the reflection event are related by

$$t' - x'/c = k(t - x/c)$$
$$t' + \frac{x'}{c} = \frac{1}{k}\left(t + \frac{x}{c}\right).$$

These may be rewritten

$$t' = \frac{1}{2}\left(k + \frac{1}{k}\right)\left[1 - \frac{vx}{c^2}\right] = \beta\left(t - \frac{vx}{c^2}\right),$$

$$x' = \frac{1}{2}\left(k + \frac{1}{k}\right)[x - vt] = \beta(x - vt),$$

where $\quad \beta = (1 - v^2/c^2)^{-1/2}.$

In these equations the coordinate is drawn in the direction of relative motion. If the two observers are moving in some other direction the formulae will, of course, be somewhat more complicated, and the simplest way of deriving them is to notice that the results which we have obtained already leave unchanged the expression $t^2 - x^2/c^2$. Since, however, with three rectangular axes the quantities y and z are also unchanged, we can see that the expression $c^2 t^2 - x^2 - y^2 - z^2$ is left unchanged as well.

It is useful to modify the notation to take account of this new point of view. If the space and time coordinates are considered together, by defining

$$x^0 = ct, \quad x^1 = x, \quad x^2 = y, \quad x^3 = z,$$

where it is convenient to write the suffixes at the top (and no confusion results since we only need to raise quantities to powers very occasionally), the quantity kept constant by the transformations may be written as

$$\eta_{\alpha\beta} x^\alpha x^\beta,$$

where $\quad \eta_{\alpha\beta} = 0, \quad \text{if} \quad \alpha \neq \beta$
$\quad \eta_{00} = 1,$
and $\quad \eta_{11} = \eta_{22} = \eta_{33} = -1.$

Here we have adopted the *summation convention* (originated by Einstein (1916) in the paper that forms the basis of his theory of gravitation and which is reproduced in the companion volume on general relativity), that a repeated literal suffix automatically carries with it an injunction to sum over appropriate values (here $0, \ldots, 3$). This formulation is derived from an idea of Minkowski (1908), who, however, preferred a slightly different one in which ict, instead of ct, is used as a coordinate, so that the invariant quantity looks more like a sum of squares. It is now widely realised that Minkowski's formulation introduces more difficulties than it solves, and can, indeed, lead the reader into numerous wrong ideas about the structure of the transformation group.

This group of transformations is known as the Lorentz group. We can see at once that the group splits up into several parts. Firstly there will be transformations of the form $t \to -t$, $\mathbf{r} \to \mathbf{r}$ which have the effect of reversing the direction of time measurement. Such transformations will only be important as special tricks for dealing with particular physical problems, not in the case which we have been considering of observers moving relative to each other. We shall restrict ourselves to the part of the group which does not involve these transformations, the so-called *isochronous transformations*. Amongst these transformations there will then be a subset of the form $t \to t$, $\mathbf{r} \to -\mathbf{r}$ corresponding to a transformation from right-handed to left-handed axes. These transformations we shall also disregard, so that, as far as the spatial part of the transformation is concerned, leaving aside time transformations, we shall confine ourselves to the *proper orthogonal group*.

We now have to consider the problem of how to make sure that the quantities which we deal with as a result of experiment really *are* invariant under the group of transformations in which we are interested. The technique employed for this purpose is a somewhat ingenious one. Suppose that we have two coordinate systems A and B, and that in these systems the results of a set of experiments are represented respectively by Q_A, Q_B, which are two sets of numbers. Consider the transformation $T_{AB}: A \to B$ which carries one coordinate system into the other. This transformation produces a corresponding transformation amongst the sets of numbers which we may denote by $T^*_{AB}: Q_A \to Q_B$. Suppose, now, that there is a third coordinate system C; then it is evidently a matter of indifference whether we transform directly from A to C or whether we go intermediately through the system B, so that the transformations of the coordinates are related by $T_{AC} = T_{AB}T_{BC}$ where the product represents the application of the transformations in succession (the left-hand one first).

In order that the number which we have been calculating shall be genuine properties of the physical system, and not merely of the coordinates in terms of which the system is described, it is then necessary that the corresponding transformations of the numbers should satisfy $T^*_{AC} = T^*_{AB}T^*_{BC}$. This is put technically in the form that the transformations of the numbers should be a *representation* of the original transformation group.

There is a technique for discovering representations of groups of coordinate transformations which can be described in general terms as follows. One looks for certain quantities which one knows on other grounds to be objects independent of the coordinate-system or *geometrical objects*, and therefore transformable under a representation of the group. One investigates this representation and then generalises to any set of objects transforming under it. For example, with the coordinate transformation group $x^\alpha \to \bar{x}^\alpha = \bar{x}^\alpha(x^\beta)$ we can consider the set of quantities dx^α, which are such that

$$d\bar{x}^\alpha = \frac{\partial \bar{x}^\alpha}{\partial x^\beta} dx^\beta.$$

Since these quantities represent a unique displacement of the point, they are obviously objects transforming under a representation of the original group, so that the linear transformation which we have found must be this representation, and any other quantity

$$\bar{A}^\alpha = \frac{\partial \bar{x}^\alpha}{\partial x^\beta} A^\beta,$$

transforming under the same representation, will be called a *contravariant vector* (the fact that this transformation is a representation of the group can be verified independently by noting that

$$\frac{\partial \bar{\bar{x}}^\alpha}{\partial \bar{x}^\beta} \cdot \frac{\partial \bar{x}^\beta}{\partial x^\gamma} = \frac{\partial \bar{\bar{x}}^\alpha}{\partial x^\gamma}\Bigg).$$

The array of differential coefficients will in our case actually be an array of constants. Similarly the array of quantities arising in the expression for the gradient of an invariant, $\partial \phi / \partial x^\alpha$ transform under

$$\frac{\partial \phi}{\partial \bar{x}^\alpha} = \frac{\partial x^\beta}{\partial \bar{x}^\alpha} \frac{\partial \phi}{\partial x^\beta}$$

and are taken as the prototype of a *covariant vector*,

$$\bar{B}_\alpha = \frac{\partial x^\beta}{\partial \bar{x}^\alpha} B_\beta.$$

By the same argument, they evidently transform under a representation of the group, as do the array $A^\alpha B_\beta = C^\alpha_\beta$, *known as a tensor of rank 2*. In the same way it is obvious that higher-order tensor representations can be defined. It will be seen in Chapter 4, however, that these are not all the representations.

We defer until the next chapter the reformulation of mechanics to be consistent with transformations of the Lorentz group, which is dealt with in Einstein's paper (Extract 5 of the present book), as indeed it is in those of Lorentz and Poincaré (Extracts 3 and 4). Indeed at this point the reader who wishes to make only a preliminary study of the subject, as explained in the Introduction, may pass on to Chapter 3, taking for granted any results that depend on the theory in the ensuing paragraphs. In a complete treatment

EINSTEIN'S CONTRIBUTION 23

there is, however, one technique which occurs in a major way in Poincaré's paper, and also plays a part in Extract 2, and that is, the use of a variational principle. Both Poincaré and Lorentz based part of their arguments on a variational principle, and because of the importance of these arguments, and the use which is made of such principles later in the theory, it is necessary here to say something about them.

Consider first the motion of a particle in one dimension in a potential field and let us limit ourselves, to begin with, to the non-relativistic case, so that the equation of motion is

$$m\ddot{x} = -\frac{\partial V}{\partial x}.$$

As is well known, there is a first integral of this, the energy integral,

$$\tfrac{1}{2}m\dot{x}^2 + V = E,$$

or

$$T + V = E.$$

This result suffices to tell us everything about the motion. If, however, the motion had been one with two or more degrees of freedom we would have had here only one first integral, and we would have had the problem of finding other results in order to determine the motion completely. Historically the approach to this problem has been strongly motivated by the experiments of physicists in a related subject, optics. Consider for a moment the phenomena of reflection and refraction of light. The law of reflection, that the angle of incidence equals the angle of reflection, could be described in slightly different terms as follows. The path of the actual ray of light from the object to the eye can be derived by joining the eye by a straight line to the mirror image of the object and then joining the intersection of this ray and the mirror to the actual object. Since a straight line is the shortest distance between two points this may also be expressed by saying that the light travels from the object to the eye by way of the mirror by the shortest path.

24 SPECIAL RELATIVITY

The problem of refraction is a little more complicated. Evidently here the path cannot be the shortest, since we have to derive not the equality between the angles, but Snell's law (Descartes, 1637). However, human ingenuity is usually able to discover some quantity which is least and Fermat (1657) discovered the appropriate quantity which was a minimum in this case, that is, the time. If the medium into which refraction takes place has a refractive index

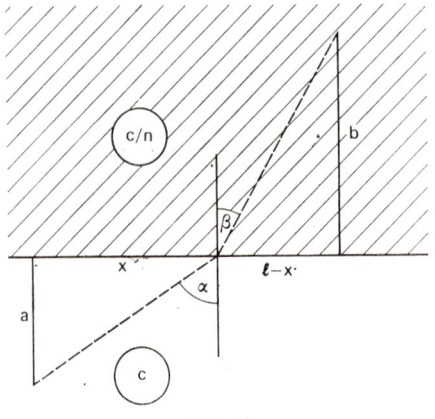

Fig. 6

n the speed of light in this medium is known to be c/n so that the expression for the time, which has to be a minimum value (Fig. 6),

$$\frac{\sqrt{(a^2+x^2)}}{c} + \frac{\sqrt{[b^2+(l-x)^2]}}{c/n}$$

gives at once $\quad \sin \alpha - n \sin \beta = 0.$

This observation is the basis of Fermat's principle of least time

$$\int n \, ds = \text{minimum}$$

and since the speed of light is unchanged by reflection at a mirror, it applies also in that case.

EINSTEIN'S CONTRIBUTION

The optical problem suggested the corresponding dynamical problem, which can be put in the form: to find a function $f = f(x, \dot{x})$ which is such that the condition

$$\int f(x, \dot{x})\, dt = \text{minimum}$$

is equivalent to the equation of motion. Now although the function f might be any function, it is natural to look first for functions which depend on invariants of the motion, and the only two invariants known so far in this simple case are the kinetic and the potential energies. This suggests that it ought to be possible to find a principle of the form

$$\int g(T, V)\, dt = \text{minimum}$$

which will be equivalent to the equations of motion. In order to find minimum values of this sort in which the quantity to be minimised depends on a curve, the path of the particle, the technique employed is very like that employed in finding the minimum value of an ordinary function; that is, one alters the argument a little and sets down the condition that the value of the function for the altered argument only differs from the original value by second-order terms. In this case the argument of the function is not a number but the particular motion from a given starting-point to a given end point. The change in the value of the quantity when this motion is slightly changed will have the form

$$\delta \int f(x, \dot{x})\, dt = \int \left(\frac{\partial f}{\partial x} \delta x + \frac{\partial f}{\partial \dot{x}} \delta \dot{x} \right) dt.$$

We can, however, simplify this form by noticing that

$$\delta \dot{x} = \frac{dx'}{dt} - \frac{dx}{dt} = \frac{d}{dt}(x' - x) = \frac{d}{dt} \delta x,$$

so that an integration by parts becomes possible and leads to

$$0 = \delta \int f(x, \dot{x}) \, dt = \int \left[\frac{\partial f}{\partial x} - \frac{d}{dt} \left(\frac{\partial f}{\partial \dot{x}} \right) \right] \delta x \, dt + \left[\frac{\partial f}{\partial \dot{x}} \delta x \right].$$

Let us then consider variations of the motion which leave the end points fixed. The second term on the right-hand side then vanishes at both limits, and the first term must therefore vanish. This must be true for arbitrary small values of the displacement, and this is only possible if the quantity in brackets vanishes at all points, that is

$$\frac{\partial f}{\partial x} - \frac{d}{dt} \left(\frac{\partial f}{\partial \dot{x}} \right) = 0.$$

This condition for a minimum is to be equivalent to the original equation of motion. Using now the particular form of the function in terms of the kinetic and potential energies, the condition becomes

$$\frac{\partial g}{\partial V} \frac{\partial V}{\partial x} - \frac{d}{dt} \left(\frac{\partial g}{\partial T} m\dot{x} \right) = 0,$$

and it is at once obvious that this agrees with the original equation of motion so long as

$$\frac{\partial g}{\partial T} = -\frac{\partial g}{\partial V} = \text{constant}.$$

Without loss of generality we can take

$$g = L = T - V.$$

In exactly the same way when there is a number of degrees of freedom the two invariants will, in general, have the forms

$$T = \tfrac{1}{2} a_{\alpha\beta} \dot{q}^\alpha \dot{q}^\beta, \quad V = f(q^\alpha)$$

with an appropriate summation over α and β and an exactly similar argument will finish with the set of *Euler–Lagrange equations*

(Lagrange, 1788)
$$\frac{\partial L}{\partial q^\alpha} - \frac{d}{dt}\left(\frac{\partial L}{\partial \dot{q}^\alpha}\right) = 0.$$

A further generalisation of this theory is also of the greatest importance in what follows, that is the application to a *field*. By a field is meant an association of a set of quantities, the field components Q^A, say, with every point of space, that is

$$Q^A = Q^A(x^\alpha).$$

The generalisation here has the following form, that instead of the generalised coordinates of the mechanical system, we now have the field components, and what is much more important, instead of the single independent variable, the time in mechanics, we now have the *four* independent variables x^α. The corresponding variational principle therefore takes the form

$$\delta \int L\, dx^1\, dx^2\, dx^3\, dx^4 \equiv \delta \int L\, d^4x = 0,$$

and an argument exactly like the one in mechanics leads to the variational equations

$$L_A = \frac{\partial L}{\partial Q^A} - \frac{\partial}{\partial x^\mu}\left(\frac{\partial L}{\partial Q^A{}_{,\mu}}\right) = 0, \quad \left(Q^A{}_{,\mu} = \frac{\partial Q^A}{\partial x^\mu}\right).$$

As an example we could consider the case of a single scalar field and a Lagrangian

$$L = \alpha Q^2 + \beta Q^\mu Q_{,\mu}, \quad Q^\mu = \eta^{\mu\nu} Q_{,\nu}.$$

The variational condition then leads to the familiar equations

$$\alpha Q - \beta \Box Q = 0,$$

where

$$\Box Q = \eta^{\alpha\beta} Q_{,\alpha\beta}.$$

Probably the most important consequence for us of the existence of variational principles of this kind is the theorem of

Noether (1918) about invariance under continuous groups of transformations. Consider first the mechanical system in which, by the same argument as before, we have the equation

$$\delta \int L\, dt \equiv \int \delta q^\alpha L_\alpha\, dt + [p_\alpha\, \delta q^\alpha]$$

for an arbitrary variation in the generalised coordinates, *not* necessarily keeping the end points fixed. Here we employ the shorthand form

$$L_\alpha \equiv \frac{\partial L}{\partial q^\alpha} - \frac{d}{dt}\left(\frac{\partial L}{\partial \dot q^\alpha}\right) = 0$$

for the equations of motion. We now wish to distinguish between displacements of the kind which we have been considering before, which leave the time unchanged, and general displacements where the time is changed at the same time as the other coordinates. We shall use the notation $\bar\delta q^\alpha$ for a displacement in which the time is unchanged, so that the result of the variation should more properly be written as $\bar\delta \int L\, dt$. In a general displacement the change in the coordinates will be given by

$$\delta q^\alpha = \bar\delta q^\alpha + \dot q^\alpha\, \delta t,$$

and we shall also have

$$\delta \int L\, dt = \bar\delta \int L\, dt + [L\, \delta t]$$

where the quantity in square brackets comes about like this:

$$\delta \int L\, dt = \bar\delta \int L\, dt + A,$$

where A is the result of varying t only. If $t \to t+h$,

$$A = \int_{t_0+h}^{t_1+h} L(t+h)\, d(t+h) - \int_{t_0}^{t_1} L(t)\, dt$$

can be reduced, by a change of variable in the first integral, $\tau = t+h$, to

$$A = \int_{t_0+h}^{t_1+h} L(\tau)\,d\tau - \int_{t_0}^{t_1} L(t)\,dt,$$

and for small h this gives, to the first order,

$$A = h(t_1)L(t_1) - h(t_0)L(t_0) = \left[L(t)\,\delta t\right]_{t_0}^{t_1}.$$

Putting these results together, and expressing some of the time-invariant variations in terms of general variations gives the result

$$\delta \int L\,dt = \int \bar{\delta}q^\alpha L_\alpha\,dt + [p_\alpha\,\delta q^\alpha - p_\alpha \dot{q}^\alpha\,\delta t] + [L\,\delta t]$$
$$= \int \bar{\delta}q^\alpha L_\alpha\,dt + [p_\alpha\,\delta q^\alpha - H\,\delta t],$$

where

$$H = p_\alpha \dot{q}^\alpha - L.$$

The quantity H, when expressed in terms of the position co-ordinates q^α and the momenta p_α (*defined* as $\partial L/\partial \dot{q}^\alpha$) only (an expression which is always possible in classical mechanics in which L has a part, T, which is a positive definite quadratic function of the $\dot{q}^\alpha$, allowing the $\dot{q}^\alpha$ to be eliminated), is called the *Hamiltonian*. It serves a useful purpose in a different formulation of the equations of motion, since for a small variation

$$\delta H = (\delta p_\alpha)\,\dot{q}^\alpha + p_\alpha\,\delta \dot{q}^\alpha - \frac{\partial L}{\partial q^\alpha}\,\delta q^\alpha - \frac{\partial L}{\partial \dot{q}^\alpha}\,\delta \dot{q}^\alpha,$$

and using the definition of the momenta p_α gives

$$\delta H = \dot{q}^\alpha\,\delta p_\alpha - \dot{p}_\alpha\,\delta q^\alpha,$$

by using the equations of motion in the form

$$\dot{p}_\alpha = \frac{dp_\alpha}{dt} = \frac{\partial L}{\partial q^\alpha}.$$

Hence, since H is a function of p_α, q^α only,

$$\frac{\partial H}{\partial p_\alpha} = \frac{dq^\alpha}{dt}, \quad \frac{\partial H}{\partial q^\alpha} = -\frac{dp}{dt}$$

the *Hamiltonian form* of the equations of motion.

The consequences of this theory are most clearly seen when the function L has, as a symmetry (see below), some continuous group of transformations

$$q^\alpha \to \bar{q}^\alpha = f^\alpha(q^\alpha, a^i),$$

where the a^i are the parameters of the group. Since the transformations form a group there is an identity transformation and there is no loss of generality in assuming it to correspond to $a^i = 0$. For small values of a^i, then, since the group is a continuous one,

$$\bar{q}^\alpha = q^\alpha + a^i f^\alpha{}_{,i} + \dots$$

or $\qquad \delta q^\alpha = \xi^\alpha \quad$ (say).

Similarly $\qquad \bar{\delta} t = \eta \quad$ (say),

and so $\qquad \delta q^\alpha = \xi^\alpha - \eta \dot{q}^\alpha.$

To say that L has the group as a symmetry means, *not* that

$$\bar{L}(\bar{q}, \dot{\bar{q}}) = L(q, \dot{q})$$

(which is evidently always true for any L which is an invariant, as it must be for the integral to have a meaning), but that

$$\bar{L}(q, \dot{q}) = L(q, \dot{q}).$$

When a transformation is a symmetry, it is obvious that the infinitesimal variations δq^α generated by it must make

$$\delta \int L \, dt = 0$$

identically, irrespective of the equations of motion. In that case

$$\int (\xi^\alpha - \eta \dot{q}^\alpha) L_\alpha \, dt + [p_\alpha \xi^\alpha - \eta H] = 0.$$

Suppose now that the equations of motion $L_\alpha = 0$ *are* satisfied, so that also

$$[p_\alpha \xi^\alpha - \eta H] = 0.$$

The quantity in brackets is then the same at the beginning and end of the motion, and these were at arbitrary times, so it is conserved.

For example, consider the case where the Lagrangian does not contain the time explicitly so that it is invariant under the group of transformations $t \to t+\eta$ where η is constant. Inserting the values of the displacements we then have

$$[-H\eta] = 0,$$

that is to say $H = $ constant, showing that the consequence of the time invariance is the conservation of the Hamiltonian, which is classically the energy. (The reader may verify that, if $L = T-V$ as above, then $H = T+V$.)

In the same way, if we have a Lagrange function expressed by cartesian coordinates in which one coordinate, say x, is not involved, the continuous group $x \to x+\lambda$ is a symmetry where λ is constant, so that in the notation of the theorem

$$\xi^\alpha = (1, 0, 0; 0), \quad \eta = 0.$$

Inserting these values gives us

$$[p_1] = 0$$

so that the absence of the x-coordinate leads to conservation of linear momentum in the x-direction. In the same way rotational invariance, for example invariance under the group whose infinitesimal members are

$$x' = x\cos\theta + y\sin\theta \simeq x+y\theta$$
$$y' = -x\sin\theta + y\sin\theta \simeq y-x\theta,$$

leads to

$$\delta x = y\theta \qquad \delta y = -x\theta.$$

This then gives, on substitution,

$$[\theta(yp_1 - xp_2)] = 0,$$

that is, conservation of the z-component of angular momentum.

An exactly similar theory holds in the case of a field, where we have the corresponding variational formula

$$\delta \int L \, d^4x \equiv \int \delta Q^A L_A \, d^4x + \int [P_A^\mu \delta Q^A - H_\nu^\mu \delta x^\nu] \, d^3 S_\mu$$

where

$$P_A^\mu = \frac{\partial L}{\partial Q^A{}_{,\mu}}$$

and

$$H_\nu^\mu = P_A^\mu Q^A{}_{,\nu} - \delta_\nu^\mu L.$$

The particular case of great interest for the field is that in which the whole theory is Lorentz invariant. The Lagrangian then will be invariant under the group of transformations

$$x^\alpha \to x^{\alpha'} = l_\alpha^{\alpha'} x^\alpha$$

of which the infinitesimal transformations have the form

$$x^{\alpha'} \to x^\alpha = (\delta_\alpha^{\alpha'} + \varepsilon_\alpha^{\alpha'}) x^\alpha.$$

The Lorentz group is such that

$$\eta_{\alpha'\beta'} x^{\alpha'} x^{\beta'} = \eta_{\alpha\beta} x^\alpha x^\beta,$$

so that

$$\eta_{\alpha'\beta'}(\delta_\alpha^{\alpha'} + \varepsilon_\alpha^{\alpha'})(\delta_\beta^{\beta'} + \varepsilon_\beta^{\beta'}) = \eta_{\alpha\beta},$$

i.e.

$$\varepsilon_{\beta\alpha} + \varepsilon_{\alpha\beta} = 0$$

where

$$\varepsilon_{\beta\alpha} = \eta_{\mu\beta} \varepsilon_\alpha^\mu.$$

Further progress in considering the theorem in this case can however only be made when we have decided on the kind of field which we are considering because we need to know under what representation of this group the field quantities transform. We shall find it convenient to defer this until the more detailed consideration of representations in Chapter 4. The present discussion, however, will serve the reader in considering the second section of Poincarés' paper in which he sets up a variational principle (following Lorentz) in order to use it to establish the Lorentz invariance of his theory in Section 3.

CHAPTER 3

Elementary Consequences of the Lorentz Transformation

THE development of special relativity falls into three parts; the first two chapters of the present book have dealt with the first of these—the accumulation of problems, mostly about optics, by the end of the nineteenth century and their resolution by the joint labours of Lorentz, Poincaré and, particularly, Einstein. The second part consists of the further application of the theory in all the other fields in which its utility became apparent, and the consequent possibility of further and repeated experimental checks, all of which it has, so far, weathered successfully. The third part is formed by the considerable extension and further development of the theoretical foundation which was needed to apply it to quantum mechanics, and this will be dealt with in Chapter 4.

One striking result is the removal of a paradox associated with Bradley's (1728) expression for *aberration*. This phenomenon, Euler's consideration of which is described in Chapter 1, is the apparent deviation of the angle at which the stars are seen, about a mean value, as a result of the orbital velocity of the earth round the sun. To describe this, consider a source of light S at rest in a coordinate-system O' (Fig. 7); if the light is monochromatic it may be written approximately as a plane wave proportional to

$$e^{i\omega'\left[t' + \frac{x'\cos\theta' + y'\sin\theta'}{c}\right]}.$$

Referred to a coordinate-system O, however, the wave is proportional to $e^{i\omega\left[t + \frac{x\cos\theta + y\sin\theta}{c}\right]}$. The exponent must be unchanged under

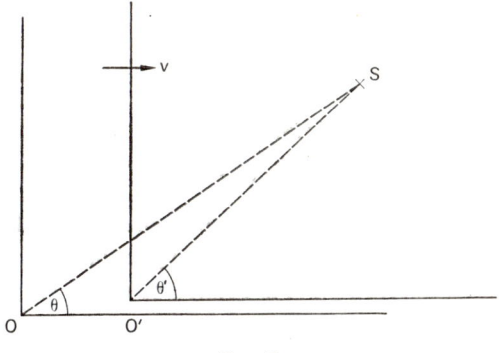

Fig. 7

the transformation, and so, using

$$x' = \beta(x-vt), \quad t' = \beta(t-vx/c^2),$$

we get

$$\beta\omega'\left[t-vx/c^2 + \frac{(x-vt)\cos\theta' + y/\beta \sin\theta'}{c}\right] = \omega\left[t + \frac{x\cos\theta + y\sin\theta}{c}\right].$$

Equating coefficients of t, x, y gives

$$\beta\omega'\left[1 - \frac{v\cos\theta'}{c}\right] = \omega,$$
$$\beta\omega'[\cos\theta' - v/c] = \omega\cos\theta,$$
$$\omega'\sin\theta' = \omega\sin\theta,$$

so that

$$\cos\theta = \frac{\cos\theta' - v/c}{1 - v\cos\theta'/c},$$

and

$$\sin\theta = \frac{\sin\theta'}{\beta[1 - v\cos\theta'/c]}.$$

The correctness of these formulae may be seen from the fact that for a source S receding directly from O, with $\theta' = 0$, it follows

that $\theta = 0$ and then
$$\omega = \beta\omega'(1-v/c) = \omega'\sqrt{\left(\frac{1-v/c}{1+v/c}\right)}.$$

Firstly, this result is *approximately*
$$\omega = \omega'(1-v/c),$$
the classical expression for the Doppler shift. Secondly, it agrees exactly with that found in Chapter 2 by means of the k-factor.

The angle of aberration, ε, is defined as $\theta - \theta'$, so that the first formula gives, since ε is small (about $41''$),
$$\cos(\theta' + \varepsilon) = \frac{\cos\theta' - v/c}{1 - v\cos\theta'/c} = \cos\theta' - \varepsilon\sin\theta',$$
so that
$$\varepsilon = \frac{v\sin\theta'/c}{1 - v\cos\theta'/c} = \beta v \sin\theta/c.$$

The classical expression of Bradley is $v\sin\theta/c$, as is to be expected. However, this expression gave rise to a paradoxical situation since an alternative derivation of it can be given in which attention is directed towards the passage of light through the telescope, which is in motion relative to the star. The motion of the telescope gives rise to the same formula, in an obvious manner; but this new derivation suggests that, if the telescope were filled with water, with velocity of light c/n, the angle of aberration should be increased by a factor n. This is not observed; so that the calculation above must be the correct one. It is instructive for the reader to use the Lorentz transformation to reproduce this argument in the coordinate system O', so as to see why the alternative argument is incorrect.

It is now time to return to the effect of Fresnel described in Chapter 1, of which a preliminary explanation was given in Chapter 2. The theory given there deals with light of one particular frequency. It is very interesting to look into the question of dispersion, that is, the different behaviour of light of different

CONSEQUENCES OF THE LORENTZ TRANSFORMATION

wavelengths. This was considered very carefully by Zeeman (1914) who repeated Fizeau's experiment with light of different colours (Extract 7 of the present book). His conclusion is that the formula given by Lorentz in 1895 (equation (3) of Zeeman's paper) is the correct one. This formula can be derived from the principles of special relativity in the following way:

The starting-point is the realisation that the speed of light in a medium which is at rest in a frame of reference O' is

$$c' = c/n(\omega')$$

where $n(\omega')$ is the refractive index expressed as a function of the frequency, ω', in the rest-frame.

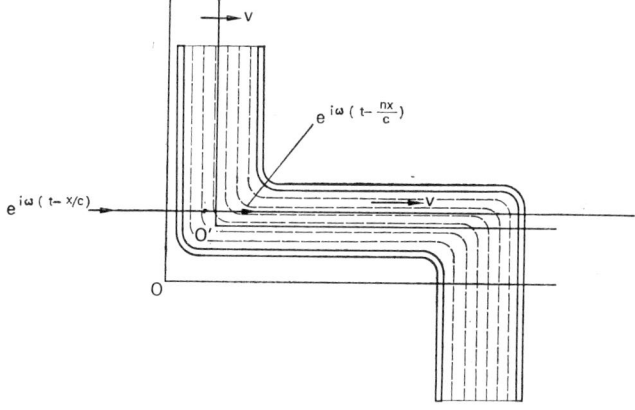

Fig. 8

In the experiment in which water flows through a pipe (Fig. 8) the light enters a medium which is *at rest relative to O*, is then carried along, and leaves similarly. Accordingly the boundary condition is that ω is unchanged. Considering, however, an incident plane wave $e^{i\omega(t-x/c)}$ which becomes $e^{i\omega'(t'-x'/c)}$ by transformation, it follows that it becomes $e^{i\omega(t-nx/c)}$ after entry into the water, the edge of which is stationary. Transforming to the axes moving with the water

$$\omega\left(t-\frac{nx}{c}\right) = \omega'(t'-x'/c') \quad \text{(say)}$$
$$= \beta\omega'\left(t-\frac{vx}{c^2}-\frac{x-vt}{c'}\right)$$
$$= \beta\omega'\left(t-\frac{nx}{c}\right)(1+v/c')$$

so long as we choose c' so that

$$\frac{n}{c}\left(1+\frac{v}{c'}\right) = \frac{v}{c^2}+\frac{1}{c'},$$

i.e. $\qquad \dfrac{1}{c'} \simeq \dfrac{n}{c}\left[1+\dfrac{nv}{c}\left(1-\dfrac{1}{n^2}\right)\right].$

Thus, to the first order in v/c,

$$\omega' \simeq \frac{\omega}{1+nv/c} \simeq \omega - \frac{n\omega v}{c},$$

and so

$$\frac{1}{n'} = \frac{1}{n} - \frac{n\omega v}{c}\frac{d}{d\omega}\frac{1}{n} = \frac{1}{n} + \frac{\omega v}{nc}\frac{dn}{d\omega}.$$

However, the measured velocity will be, as before,

$$u = \frac{c/n'+v}{1+v/n'c} \simeq \frac{c}{n'}+v\left(1-\frac{1}{n'^2}\right)$$
$$\simeq \frac{c}{n}+v\left(1-\frac{1}{n^2}+\frac{\omega v}{n}\frac{dn}{d\omega}\right).$$

Since $\omega\lambda = 2\pi c$, we can replace the last term by $-(\lambda/n)(dn/d\lambda)$, which gives the Lorentz formula quoted by Zeeman.

Of course, if the moving medium is a solid (e.g. a block of glass) the formula is a different one since the boundary condition is now that ω' is unchanged in entering and leaving, although ω is the measured frequency. Using the formula

$$\omega' \simeq \omega(1-v/c),$$

so that
$$\frac{1}{n(\omega')} = \frac{1}{n} - \frac{\omega v}{c} \frac{d}{d\omega} \frac{1}{n} = \frac{1}{n} + \frac{\omega v}{cn^2} \frac{dn}{d\omega},$$

it then follows that
$$u \simeq \frac{c}{n} + v\left(1 - \frac{1}{n^2} + \frac{\omega}{n^2} \frac{dn}{d\omega}\right)$$
$$= \frac{c}{n} + v\left(1 - \frac{1}{n^2} - \frac{\lambda}{n^2} \frac{dn}{d\lambda}\right).$$

The difference between these formulae has been emphasised by Landsberg (1961).

Coming now to investigations needing a detailed knowledge of the transformation of Maxwell's equations, we select for special consideration that of Wilson and Wilson (Extract 6 of the present book). Before 1905 electromagnetic theory had found great difficulty in dealing with moving materials, and this difficulty is particularly apparent when a magnetic dielectric (in the Wilson–Wilson experiment, sealing-wax with embedded steel balls) is under consideration. Consider, for simplicity, an infinite parallel-plate condenser with a magnetic dielectric which is moving as indicated (Fig. 9) in a magnetic field, the plates being short-circuited by brushes and a wire containing a ballistic galvanometer which are at rest relative to the magnetic field.

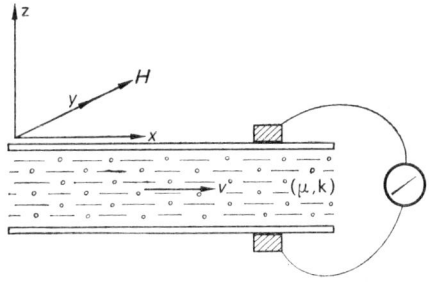

Fig. 9

Since
$$\operatorname{curl} \mathbf{E} = -\mu \frac{\partial \mathbf{H}}{\partial t} = 0,$$

it follows that $\oint \mathbf{E} \cdot d\mathbf{r} = 0$ round any closed circuit. For the wire $\mathbf{E} = 0$, so that, in the dielectric $E_3 = 0$. The constitutive relations are
$$\mathbf{D}' = \varkappa \mathbf{E}', \quad \mathbf{B}' = \mu \mathbf{H}'.$$

Now use the formulae for transformation given in Einstein's paper (Extract 5):

$$E_1' = E_1, \qquad\qquad B_1' = B_1,$$
$$E_2' = \beta\left(E_2 - \frac{vB_3}{c}\right), \quad B_2' = \beta\left(B_2 + \frac{vE_3}{c}\right),$$
$$E_3' = \beta\left(E_3 + \frac{vB_2}{c}\right), \quad B_3' = \beta\left(B_3 - \frac{vE_2}{c}\right),$$

together with the analogous results

$$H_1' = H_1, \qquad\qquad D_1' = D_1,$$
$$H_2' = \beta\left(H_2 + \frac{vD_3}{c}\right), \quad D_2' = \beta\left(D_2 - \frac{vH_3}{c}\right),$$
$$H_3' = \beta\left(H_3 - \frac{vD_2}{c}\right), \quad D_3' = \beta\left(D_3 + \frac{vH_2}{c}\right).$$

Here $H_1 = H_3 = E_3 = 0$, so that

$$D_1 = 0,$$
$$\beta D_2 = -\beta \varkappa v B_3/c,$$
$$\beta\left(D_3 + \frac{vH_2}{c}\right) = \beta\varkappa\left(\frac{vB_2}{c}\right),$$
$$B_1 = 0,$$
$$\beta B_2 = \beta\mu\left(H_2 + \frac{vD_3}{c}\right),$$
$$\beta\left(B_3 - \frac{vE_2}{c}\right) = -\beta\mu\left(\frac{vD_2}{c}\right).$$

CONSEQUENCES OF THE LORENTZ TRANSFORMATION 41

Our interest is in D_3, since the field **D** terminates at the brushes, and the relevant equations, eliminating B_2, give

$$\frac{v\varkappa\mu}{c}\left(H_2 + \frac{vD_3}{c}\right) = D_3 + \frac{vH_2}{c},$$

i.e.
$$D_3 = \frac{v}{c}\frac{\varkappa\mu - 1}{1 - v^2\varkappa\mu/c^2} H_2.$$

Reversal of the direction of **H** will therefore reverse D_3 causing a flow of charge through the ballistic galvanometer. In the actual experiment, described in Extract 6, a rotating cylinder was used as the condenser, but the principle is the same.

The problem of rendering mechanics consistent with the transformations which have been found above and so of settling the problem of inertial frames falls into two parts. The first part, with which alone we shall be concerned in the present book, consists of reformulating Newton's laws of motion to be consistent with these transformations. The second part then consists of incorporating in these laws of motion the particular field of force observed in the gravitational field. This latter problem occupied Einstein for 10 years, and led him to the general theory of relativity, which is described in the companion volume.

The transformation which has been found between inertial frames has the form

$$x' = \beta(x - vt), \quad y' = y,$$
$$t' = \beta(t - vx/c^2), \quad z' = z.$$

As a result, the velocity of a particle in the x-direction transforms under

$$u \to u' = \frac{dx'}{dt'} = \frac{u - v}{1 - uv/c^2}$$

(although the components of velocity in the y and z directions will transform in a different way). This result is, of course, consistent with the formula above accounting for the Fresnel experiment. In mechanics, however, we are concerned not with velocities but with accelerations, and when we carry out another differentiation we find

$$\frac{du'}{dt'} = \frac{dt}{dt'} \left[\frac{du/dt}{1-uv/c^2} + \frac{1}{c^2} \frac{(u-v)v(du/dt)}{(1-uv/c^2)^2} \right],$$

a result which can be further simplified by noticing that

$$\frac{dt}{dt'} = \frac{1}{\beta(1-uv/c^2)}.$$

One consequence of these equations is at once apparent; that is, that the condition

$$\frac{du}{dt} = \text{constant}$$

in one coordinate system does not lead to, and indeed is inconsistent with,

$$\frac{du'}{dt'} = \text{constant}.$$

It is clear then that the physical quantity acceleration, which in Newtonian mechanics is defined as

$$\frac{d^2x}{dt^2} = \frac{du}{dt},$$

is more appropriately defined differently in special relativity, although the two definitions ought to agree when the velocities which enter are small. Since they are to agree approximately for small velocities, the appropriate definition is fairly obvious. We should make the agreement perfect when the velocity falls to zero.

The acceleration should, then, be derived from the Newtonian measure of the acceleration in that particular coordinate system in which the particle is instantaneously at rest (say, at the begin-

CONSEQUENCES OF THE LORENTZ TRANSFORMATION

ning of the moment in which the differentiation is carried out). In other words, we should define for the acceleration in any coordinate system the value given by the above formula when $u = v$. Making this substitution we get

$$\frac{du'}{dt'} = \frac{dt}{dt'} \frac{1-v^2/c^2}{(1-uv/c^2)^2} \frac{du}{dt} = \frac{1}{\beta^3} \frac{1}{(1-uv/c^2)^3} \frac{du}{dt},$$

so that when $u = v$,

$$\frac{du'}{dt'} = \beta^3 \frac{du}{dt}.$$

At first sight such a definition seems quite in conflict with Newton's laws of motion, but this turns out not to be the case, for the following reason. Firstly, we have from the definition,

$$\frac{1}{\beta^2} + \frac{u^2}{c^2} = 1$$

and differentiating this gives us

$$-\frac{2}{\beta^3} \frac{d\beta}{dt} + \frac{2u}{c^2} \frac{du}{dt} = 0.$$

As a result it follows that

$$\frac{d}{dt}(\beta u) = \beta \frac{du}{dt} + \beta^3 \frac{u^2}{c^2} \frac{du}{dt} = \beta^3 \frac{du}{dt}.$$

Since the final result here is exactly the quantity to be defined as the acceleration we can, by inserting the mass, rewrite Newton's second law in the form

$$\frac{d}{dt}(m\beta u) = F$$

which certainly agrees with the Newtonian form when the velocities are small, and has the advantage of being true in any coordinate system. Moreover, this law is such that a constant force in one coordinate system corresponds to a constant force in another. In the

44 SPECIAL RELATIVITY

above derivation the term mass has not been clearly defined. When we included the mass in the equation in order to derive Newton's law, the number in question was intended to represent what is called in elementary work the "quantity of matter in the body," that is a constant number to be attached to the particle. If, however, we use Newton's law as modified here to *define* and measure masses, say by collision experiments (see the paper by Einstein, Extract 5, section 10), the measured mass will not be the quantity m but the coefficient of v in the expression for the momentum; thus the measured mass will be

$$m' = \beta m = \frac{m}{(1-v^2/c^2)^{1/2}}$$

and so will be found to increase with the velocity. An increase of mass with velocity had already been found experimentally for electrons before the advent of relativity (Kaufmann, 1902).

Let us now try to understand a little more fully what this increase in mass with velocity represents. Imagine first that the speeds are sufficiently small for powers of the velocity above the second to be neglected. The formula for the measured mass then takes the form

$$m' \simeq m + \tfrac{1}{2} mv^2/c^2$$

so that to the rest mass is added the Newtonian kinetic energy. This suggests firstly that the correct formulation of kinetic energy in relativity is not the Newtonian form but the different formula:

$$\text{Kinetic energy } T = (m'-m)c^2 = mc^2\left[\frac{1}{\sqrt{(1-v^2/c^2)}} - 1\right].$$

If this is so, it also suggests that the term which is subtracted here must represent some residual, present even when the velocity is zero, that is, when we study the particle from the point of view of its rest-frame. This has been called the rest energy of the particle E and we have the result $E = mc^2$ relating energy and mass.

At this point the reader making a preliminary study of the subject may pass to Chapter 4. The remainder of this chapter relies on the arguments used at the end of Chapter 2. The investigation which has just been given is suitable for the dynamics of a single particle. In classical mechanics, however, more general systems than a single particle are considered. Moreover, in quantum mechanics (which will be dealt with more fully in Chapter 4) entities arise which exhibit particle-like qualities, but other qualities as well. The single method appropriate to both of these generalisations is to express the equations of mechanics in terms of a variational principle, as in Chapter 2.

We hold fast, then, to the variational principle

$$\delta \int L \, dt = 0,$$

although the function L may have to be defined in a much more complicated manner than hitherto, and various other alterations may have to be made. In the original form the variations are those keeping the end-points fixed and the time unvaried, as was found in Chapter 2, but this fails to meet one obvious desideratum—even if L is so chosen that the resulting equations are Lorentz-invariant (which is possible), this invariance is concealed by the fact that t enters in a completely different manner from the space-coordinates.

Consider, for example, the Lagrangian

$$L = [m_0 + \phi(\mathbf{r})/c^2](1 - \mathbf{v}^2/c^2)^{1/2}.$$

The Euler–Lagrangian equations are

$$\frac{1}{\beta}\nabla\phi + \frac{d}{dt}\left[\frac{(m_0+\phi/c^2)\mathbf{v}}{\sqrt{(1-v^2/c^2)}}\right] = 0,$$

or

$$\frac{d}{dt}[\beta(m_0+\phi/c^2)\mathbf{v}] = -\frac{1}{\beta}\nabla\phi,$$

and this is the correct relativistic equation of motion for a particle under a potential field ϕ. The mass is now

$$\beta(m_0+\phi/c^2) \simeq m_0+(\tfrac{1}{2}m_0v^2+\phi)/c^2,$$

giving, approximately, a contribution to the rest mass not only from the kinetic energy but from the potential energy as well.

Even more is this the case in the Hamiltonian (or canonical) formulation now to be described. According to the fundamental variational formula of Chapter 2, when the time is varied as well as the other coordinates,

$$\delta \int L\, dt = [p_\alpha\, \delta q^\alpha - H\, \delta t] + \int L_\alpha \bar\delta q^\alpha dt,$$

where $H = p_\alpha \dot q^\alpha - L$. The quantity H, called the *Hamiltonian*, is a useful quantity in terms of which to express the variational principle. It will turn out later, however, that in the generalisations we shall make the dual roles played by the Hamiltonian (the conserved quantity in Noether's theorem and the function from which, as we are about to show again, may be derived first-order equations of motion) must be separated. Since

$$L = p_\alpha \dot q^\alpha - H$$

the equations of motion are, in fact, given by

$$\delta \int (p_\alpha \dot q^\alpha - H)\, dt = 0,$$

or
$$\delta \int (p_\alpha\, dq^\alpha - H\, dt) = 0.$$

Performing the variation again using this form, and remembering that $d\delta t = \delta dt = 0$, we have, after an integration by parts,

$$\int (\delta p_\alpha\, dq^\alpha - \bar\delta q^\alpha\, dp_\alpha - \delta H\, dt) = 0,$$

so

$$\delta H = \dot{\delta p}_\alpha \frac{dq^\alpha}{dt} - \delta q^\alpha \frac{dp_\alpha}{dt}.$$

It follows that, if H is treated as a function of p_α and q^α,

$$\dot{q}^\alpha = \frac{\partial H}{\partial p_\alpha}, \quad \dot{p}_\alpha = -\frac{\partial H}{\partial q^\alpha},$$

the Hamiltonian form of the equations of motion, as derived in Chapter 2 (Hamilton, 1835).

It is useful to look at these equations as relating the coordinates of a point of a $2n$-dimensional space, by defining

$$X^A = \{q^1, q^2, \ldots, q^n, p_1, p_2, \ldots, p_n\}, \quad (A = 1, \ldots, 2n)$$

so that

$$\frac{dX^A}{dt} = \theta^{AB} \frac{\partial H}{\partial X^B},$$

where $\quad \theta^{\alpha\, n+\alpha} = 1 \quad \theta^{n+\alpha\, \alpha} = -1$

and $\quad \theta^{AB} = 0$ otherwise.

In considering the transformation of such equations to a new system of X-coordinates, the tensor rule gives

$$\frac{dX^{A'}}{dt} = \theta^{A'B'} \frac{\partial H}{\partial X^{B'}}$$

where $\theta^{A'B'}$ is the transform of θ^{AB},

$$\theta^{A'B'} = \frac{\partial X^{A'}}{\partial X^A} \frac{\partial X^{B'}}{\partial X^B} \theta^{AB}.$$

But this new equation will only again be the Hamiltonian equations of motion if θ^{AB} is unchanged; if, then, $\theta^{A'B'}$ is understood to be defined as $+1$ or 0 like θ^{AB}, the transformation equation for θ^{AB} becomes a differential equation for the transformation. Such differential equations are abbreviated in the usual treatment by defining the *Poisson bracket* of two function ϕ, ψ by

$$(\phi, \psi) = \theta^{AB} \frac{\partial \phi}{\partial X^A} \frac{\partial \psi}{\partial X^B}.$$

The differential equation is then $(X^{A'}, X^{B'}) = \theta^{A'B'}$. Moreover, the equation of motion of the system is able to be written

$$\frac{dX^A}{dt} = (X^A, H).$$

In fact Dirac was able to show that all the important formulae of the Hamiltonian theory could be expressed in terms of Poisson brackets.

The moral of all this is clear, however; instead of t some other parameter should be used to fix the points of the curve, leaving t free to be treated like the other coordinates. (Such a technique is not specifically special relativistic; indeed, as we said, it is not essential, but only a great convenience here. On the other hand, for the Newtonian mechanics of a system in which the Lagrangian contains the time explicitly—as, for instance, a simple pendulum whose length is constrained to vary in some way—the technique is again of great convenience.) Denoting the parameter by u, we write

$$L\, dt = f\, du$$

where f is now a function of *all* the coordinates and velocities (so of t and dt/du in general). Now since, denoting differentiation with respect to u by an accent, it is obvious that, writing

$$\grave{q}^a = \frac{dq^a}{du},$$

$$L = L(q^a, \dot{q}^a) = L\left(q^a, \frac{\grave{q}^a}{\grave{q}^0}\right) \qquad (a, b = 0, \ldots, n, \quad q^0 = t)$$

is a function of the ratios of the new velocities only, i.e. it is homogeneous of degree zero in these new velocities, and so $f = L\grave{q}^0$ is homogeneous of degree one in these velocities. A general variation gives

$$\delta \int f\,du = \int \left(\frac{\partial f}{\partial q^a}\delta q^a + \frac{\partial f}{\partial \grave{q}^a}\delta \grave{q}^a \right) du$$

$$= \int \left(\frac{\partial f}{\partial q^a} - \frac{d}{du}\frac{\partial f}{\partial \grave{q}^a} \right) \delta q^a + \left[\frac{\partial f}{\partial \grave{q}^a}\delta q^a \right].$$

But $\delta \int f\,du = \delta \int L\,dt$, so equating terms gives us

$$p_\alpha = \frac{\partial f}{\partial \grave{q}^\alpha}, \qquad\qquad H = -\frac{\partial f}{\partial \grave{q}^0},$$

$$L_\alpha = \frac{\partial f}{\partial q^\alpha} - \frac{d}{du}\left(\frac{\partial f}{\partial \grave{q}^\alpha}\right), \quad \frac{\partial f}{\partial q^0} - \frac{d}{du}\left(\frac{\partial f}{\partial \grave{q}^0}\right) = 0.$$

The apparently new equation of motion is, in fact,

$$0 = \frac{\partial}{\partial t}(L\dot{t}) - \frac{d}{du}\left[\frac{\partial}{\partial \dot{t}}(L\dot{t})\right]$$

$$= \dot{t}\left[\frac{\partial L}{\partial t} - \frac{d}{dt}\left(L - \dot{q}^\alpha \frac{\partial L}{\partial \dot{q}^\alpha}\right)\right]$$

$$= \dot{t}\left[\frac{\partial L}{\partial t} - \frac{d}{dt}(L - p_\alpha \dot{q}^\alpha)\right] = \dot{t}\left(\frac{\partial L}{\partial t} + \frac{dH}{dt}\right),$$

which is an identity that could have been deduced from the others.

The corresponding canonical formalism is now a rather more complicated matter. Let us begin by carrying through a few stages in what one would imagine to be the obvious generalisation, so as to point out the difficulties into which it runs. We can then carry out the necessary modifications at the end of this chapter. Since f is homogeneous of degree one in the $\grave{q}^a$, it is clear that the

$$p_a = \frac{\partial f}{\partial \grave{q}^a}$$

50 SPECIAL RELATIVITY

are homogeneous of degree zero in the $\dot{q}^a$, i.e. depend only on their ratios, and these n ratios can therefore be eliminated, giving, say,

$$\Omega(q^a, p_a) \equiv 0.$$

Further, by Euler's theorem in homogeneous functions,

$$p_a \dot{q}^a = \dot{q}^a \frac{\partial f}{\partial \dot{q}^a} = f$$

so that the variational principle can be rewritten

$$\delta \int f \, du = \delta \int p_a \, dq^a = 0,$$

which, after integration by parts, gives

$$\int (\delta p_a \, dq^a - dp_a \, \delta q^a) = 0$$

for arbitrary variations, subject only to

$$\Omega(q^a, p_a) = 0.$$

[Note that this formula is *like* that derived above in the form

$$\int (\delta p_\alpha \, dq^\alpha - \delta q^\alpha \, dp_\alpha - \delta H \, dt) = 0$$

since $H = -p_0$ and $t = 0$. Now, however, a general variation is permitted, and so the relation $\Omega = 0$ has to be added.] The equations of motion are therefore given by

$$\delta p_a \, dq^a - dp_a \, \delta q^a = 0$$

for any variations δp_a, δq^a such that

$$\frac{\partial \Omega}{\partial q^a} \delta q^a + \frac{\partial \Omega}{\partial p_a} \delta p_a = 0,$$

i.e. by

$$dq^a = \phi \frac{\partial \Omega}{\partial p_a}, \quad dp_a = -\phi \frac{\partial \Omega}{\partial q^a}.$$

Clearly ϕ must have the form dw, where w is some new parameter, so the canonical equations now take the form

$$\frac{dq^a}{dw} = \frac{\partial \Omega}{\partial p_a}, \quad \frac{dp_a}{dw} = -\frac{\partial \Omega}{\partial q^a}.$$

A very singular complication now arises: the quantity which, in these relations, replaces H in Hamilton's equations of motion, is, in fact, zero (for the way Ω enters is by means of the equation $\Omega(q, p) = 0$). For example, one form for Ω is

$$\Omega \equiv p_0 + H(q^\alpha, p_\alpha) = 0,$$

but this form picks out the time-coordinate and deals with it differently from the others. When this form is chosen for Ω the parameter w is t. The reason for this complication, which precludes Ω being identified with any kind of energy, is essentially the following: if the generalised momenta are defined by

$$p_a = \frac{\partial f}{\partial \dot{q}^a},$$

they are homogeneous of degree zero in the $\dot{q}^a$ and therefore (by Euler's theorem)

$$\dot{q}^a \frac{\partial p_b}{\partial \dot{q}^a} = 0,$$

i.e.

$$\dot{q}^a \frac{\partial^2 f}{\partial \dot{q}^a \partial \dot{q}^b} = 0.$$

It follows that the determinant

$$\left| \frac{\partial^2 f}{\partial \dot{q}^a \partial \dot{q}^b} \right| \equiv 0,$$

i.e.

$$\left| \frac{\partial p_b}{\partial \dot{q}^a} \right| = 0,$$

and this equation implies that we cannot solve for the $\dot{q}^a$ in terms of the p_b. (Of course, since only their ratios are involved!) The dynamical theory is therefore of the kind called *singular*.

Two possibilities now arise: one is to learn to work with the above formalism, admitting that the corresponding classical case is only going to serve as a very rough guide indeed. The other, which has been preferred both in general relativity, which, as we shall see in the companion volume, is a field theory which is singular in the same way, and in relativistic quantum mechanics, is to take the view that this singularity of the theory is an indication that p_a, q^a are not, after all, a suitable set of variables in terms of which to construct the canonical theory, and to replace the p_a by some new variables, π_a say, which are more suitable (see Rund, 1966). It is necessary to explain this at some length, so that the application to quantum mechanics can be made clear in the next chapter. The new approach starts from the remark that, for an ordinary dynamical system in which

$$L = \tfrac{1}{2} a_{\alpha\beta} \dot{q}^\alpha \dot{q}^\beta - V,$$

the matrix
$$\frac{\partial^2 L}{\partial \dot{q}^\alpha \, \partial \dot{q}^\beta} = a_{\alpha\beta},$$

so that a singular theory cannot arise in Newtonian mechanics, and, when it does arise, it is associated with an unusual form of kinetic energy. Now, in general, when, for *any* function f of variables the determinant

$$\left| \frac{\partial^2 f}{\partial \dot{q}^a \, \partial \dot{q}^b} \right| = 0,$$

it will *not* be the case that

$$\left| \frac{\partial^2 f^2}{\partial \dot{q}^a \, \partial \dot{q}^b} \right| = 0$$

as well. (For example, in the simplest case, in which there is one variable only, so that

$$\frac{\partial^2 f}{\partial \dot{q}^2} = 0, \quad \text{implying} \quad f = a\dot{q} + b,$$

where a, b are functions of q only, we have

$$\frac{\partial^2 f^2}{\partial \dot{q}^2} = \frac{\partial}{\partial \dot{q}}\left(2f\frac{\partial f}{\partial \dot{q}}\right) = 2a^2,$$

which cannot be zero if f involves $\dot{q}$ at all.) Let us make the assumption, then, that

$$\left|\frac{\partial^2 f^2}{\partial \dot{q}^a \partial \dot{q}^b}\right| \not\equiv 0,$$

and let us proceed to set up a canonical theory in terms of the variable q^a and

$$\pi_a = \frac{\partial}{\partial \dot{q}^a}\left(\frac{1}{2}f^2\right)$$

(where the factor $\frac{1}{2}$ has been inserted for convenience). We may notice that, since $\pi_a = fp_a$, the π_a are homogeneous of degree *one* in the $\dot{q}^a$. Moreover, since $f = p_a \dot{q}^a$, it follows that

$$\pi_a \dot{q}^a = f^2.$$

On the other hand, from our assumption, we can solve for the $\dot{q}^a$ in terms of the π_a, and in fact the $\dot{q}^a$ are also homogeneous of degree one in the π_a; for we have the obvious identity

$$\frac{\partial \dot{q}^a}{\partial \pi_b}\frac{\partial \pi_b}{\partial \dot{q}^c} = \delta^a_c,$$

so that
$$\frac{\partial \dot{q}^a}{\partial \pi_b}\frac{\partial^2(\frac{1}{2}f^2)}{\partial \dot{q}^c \partial \dot{q}^b} = \delta^a_c,$$

and therefore

$$\dot{q}^a = \delta^a_c \dot{q}^c = \dot{q}^c \frac{\partial \dot{q}^a}{\partial \pi_b} \frac{\partial^2(\tfrac{1}{2}f^2)}{\partial \dot{q}^c \, \partial \dot{q}^b}$$

$$= \frac{\partial \dot{q}^a}{\partial \pi_b} \left\{ \frac{\partial}{\partial \dot{q}^b}\left(\dot{q}^c \frac{\partial(\tfrac{1}{2}f^2)}{\partial \dot{q}^c} \right) - \delta^c_b \frac{\partial(\tfrac{1}{2}f^2)}{\partial \dot{q}^c} \right\}$$

$$= \frac{\partial \dot{q}^a}{\partial \pi_b} \frac{\partial}{\partial \dot{q}^b}\left(\frac{1}{2}f^2\right) = \pi_b \frac{\partial \dot{q}^a}{\partial \pi_b},$$

which shows $\dot{q}^a$ to be homogeneous, of degree one, in the π_b.

We now need a function $\mathcal{H} = \mathcal{H}(q^a, \pi_a)$ which will serve as a generator of the canonical equations of motion. The unexpected result of this investigation is that the same quantity, $\tfrac{1}{2}f^2$, that generates momentum variables also serves for this purpose.

An analogy with the non-relativistic case is what suggests this; for there the Hamiltonian

$$H = p_\alpha \dot{q}^\alpha - L,$$

and so here it is natural to try

$$\mathcal{H} = \pi_a \dot{q}^a - \tfrac{1}{2}f^2.$$

Since, however, $\pi_a \dot{q}^a = f^2$, this suggests trying $\mathcal{H} = \tfrac{1}{2}f^2$.

More precisely, we let

$$\mathcal{H}(q^a, \pi_a) \equiv \tfrac{1}{2}[f(q^a, \dot{q}^a)]^2$$

when we have substituted for the $\dot{q}^a$. Let us first verify that this does indeed give the canonical equations. We have, firstly, that

$$\frac{\partial \mathcal{H}}{\partial \pi_a} = \frac{\partial}{\partial \pi_a}\left(\frac{1}{2}\pi_b \dot{q}^b\right) = \frac{1}{2}\left(\dot{q}^a + \pi_b \frac{\partial \dot{q}^b}{\partial \pi_a}\right).$$

On the other hand,

$$\frac{\partial \mathcal{H}}{\partial \pi_a} = f \frac{\partial f}{\partial \pi_a} = f \frac{\partial f}{\partial \dot{q}^b} \frac{\partial \dot{q}^b}{\partial \pi_a} = \pi_b \frac{\partial \dot{q}^b}{\partial \pi_a}.$$

Comparing these results allows us to deduce that

$$\frac{\partial \mathcal{H}}{\partial \pi_a} = \frac{dq^a}{du}.$$

Consider next

$$\frac{d\pi_a}{du} = \frac{d}{du}(fp_a) = \frac{df}{du}p_a + f\frac{dp_a}{du} = \frac{df}{du}p_a + f\frac{\partial f}{\partial q^a},$$

using the equations of motion. A little care is now needed; at first sight it appears that, since $\mathcal{H} = \frac{1}{2}f^2$ the second term in this equation would be $\partial \mathcal{H}/\partial q^a$. But $\partial/\partial q^a(\frac{1}{2}f^2)$ refers to a differentiation keeping $\dot{q}^a$ constant, whilst $\partial \mathcal{H}/\partial q^a$ refers to one keeping π_a constant. We can establish the connection between these as follows:

$$\frac{\partial \mathcal{H}}{\partial q^a} = f\frac{\partial f}{\partial q^a} + f\frac{\partial f}{\partial \dot{q}^b}\frac{\partial \dot{q}^b}{\partial q^a}.$$

Since, however

$$\dot{q}^b = \frac{\partial \mathcal{H}}{\partial \pi_b},$$

we have

$$\frac{\partial \dot{q}^b}{\partial q^a} = \frac{\partial^2 \mathcal{H}}{\partial q^a\, \partial \pi_b}.$$

Accordingly, the last term becomes

$$f\frac{\partial f}{\partial \dot{q}^b}\frac{\partial^2 \mathcal{H}}{\partial q^a\, \partial \pi_b} = \pi_b\frac{\partial^2 \mathcal{H}}{\partial q^a\, \partial \pi_b} = \frac{\partial}{\partial q^a}\left(\pi_b\frac{\partial \mathcal{H}}{\partial \pi_b}\right) = 2\frac{\partial \mathcal{H}}{\partial q^a}.$$

In all, then

$$\frac{\partial \mathcal{H}}{\partial q^a} = -\frac{\partial}{\partial q^a}\left(\frac{1}{2}f^2\right),$$

and so we have the two canonical equations

$$\frac{dq^a}{du} = \frac{\partial \mathcal{H}}{\partial \pi_a}, \quad \frac{d\pi_a}{du} = -\frac{\partial \mathcal{H}}{\partial q^a} + \frac{1}{f}\frac{df}{du}\pi_a.$$

56 SPECIAL RELATIVITY

The second of these differs from the usual form; but since u is an arbitrary parameter along the path, we may replace it by a new parameter, w, say, and because

$$\mathcal{H} = \tfrac{1}{2}\pi_a \dot{q}^a$$

we may choose w so that

$$\mathcal{H}\frac{du}{dw} = 1, \quad \text{i.e.} \quad \pi_a \frac{dq^a}{dw} = 1.$$

With the new parameter, then, $\mathcal{H}$ becomes 1, so the extra term vanishes (since $\mathcal{H} = \tfrac{1}{2}f^2$).

To summarise, then, we have shown that, even in the case of a relativistic system, there *is* an alternative canonical formulation, in which the Hamiltonian is of the form $\mathcal{H} = \pi_a \dot{q}^a$. This alternative formulation is the one that corresponds to the quantum-mechanical treatment of Dirac in Extract 8, althounhg he does not say so there.

CHAPTER 4

Applications in Quantum Theory

IN ORDER to describe the more recent developments in special relativity it is necessary to say a little about the branch of physics which sparked them off, that is quantum mechanics. The general atmosphere in physics at the end of the nineteenth century was one of considerable optimism. One of the few outstanding problems which remained was the question of the interaction between radiation and matter. The behaviour of radiation away from matter was well understood, being described by Maxwell's equations. What was difficult to understand was the way in which the radiation in equilibrium with matter at a given temperature has a definite frequency distribution. Since the temperature does not enter into Maxwell's equations it is impossible to answer this problem on the basis of them alone.

A certain amount was known already by 1900 about this problem because by thermodynamical arguments Wien's displacement law (1893) showed that when the distribution of energy is known at any one temperature it can be calculated at any other temperature. We know, of course, that the total energy over the whole spectrum for temperature T is proportional to the fourth power of T (Stefan's law). Wien's law is accordingly best stated in terms of the energy for a given temperature and frequency divided by the fourth power of the temperature. More precisely, let $u(\nu, T)\,d\nu$ be the energy density of radiation with frequencies between ν and $\nu+d\nu$. Let $\nu/T = \omega$; then by general thermodynamical considerations one can prove that

$$u(\nu, T)\,d\nu = [f(\omega)\,d\omega]T^4.$$

What is left unspecified by Wien's law is *how* the square bracket depends on that ratio. Lord Rayleigh (1900), on the basis of rather tentative arguments, in which the radiation is treated as vibrations of an elastic aether derived one formula corresponding to $f(\omega) = a\omega^2$, where a is a numerical factor. Although this formula is correct at low frequencies it cannot possibly be correct for all, since the total energy in the spectrum would be infinite. Earlier Wien (1896), by a rather complicated argument using the kinetic theory of gases, had derived another formula which is experimentally correct for high frequencies but wrong for the low ones, with $f(\omega) = b\omega^3 e^{-a\omega}$, where a, b are constants.

As is well known, the mystery here was dispelled by Max Planck (1900) at the turn of the century with the formula

$$f(\omega) = \frac{8\pi h}{c^3} \frac{\omega^3}{e^{h\omega/k}-1},$$

which evidently can agree with Rayleigh's formula for small ω, and with Wien's for large ω, so long as h is not allowed to be arbitrarily small (as Planck expected) but given a definite experimental value ($6\cdot6 \times 10^{-27}$ cgs units). Planck's modified formula was interpreted as meaning that the energy of the system, which had previously been treated as a continuous variable, could not in fact be continuous but changed by steps, the size of each step being proportional to the frequency of the radiation emitted. The constant of proportionality, Planck's constant, was exceedingly small in terms of everyday measurements, thus accounting for the apparent continuity of the energy. Because of this apparent continuity energy can be treated as a continuous variable in everyday experience, and it is only in certain rather subtle and rather complex situations that it will be necessary to use the fact that it is a discrete quantity, or as is usually said, quantized.

By a fortunate chance, not unusual in the history of science, at this very time numerous situations of this subtle kind were arising in different experimental fields. J. J. Thompson (1897) had measured in Cambridge in 1894 the speed of the rays emitted by an electrical discharge through a near-vacuum. Because the speed was about a thousand times slower than that of light he formed the opinion that these rays were really streams of particles, and accordingly sought to measure whether they had a charge. Within 3 years he had determined their charge–mass ratio, and in another 12 months, with Townsend, he had proved that the charge was constant. These particles were, of course, electrons, and the discrete nature of electrical charge was thus discovered only 1 year before that of energy. The connection between these two quantizations is still a mystery, though a clue to it may perhaps lie in the fact that the ratio $hc/2\pi e^2$ is a pure number with a value close to 137. In the years at the beginning of the century, Rutherford (1911) gradually pieced together experimental conclusions about scattering, leading him to the discovery of the atomic nucleus and so to a definite atomic model. Two years later Bohr (1913) was able to apply Planck's idea to the explanation of the discrete lines in the hydrogen spectrum. This was the real beginning of the old quantum theory, though this name is slightly misleading since it is true to say that there never was any complete *theory*, worthy of the name. Rather there was a collection of rules which solve particular problems, but the use of these rules was a matter for artistry, so that only the initiated were able to tackle any new problem. Moreover, there were experimental results which it seemed that the theory was quite inadequate to deal with, such as the relative intensity of the various lines into which a given line was split by electric and magnetic fields. As the theory went on it became more and more difficult because of additional *ad hoc* assumptions, and it gradually became thought that some totally new direction had to be taken. (Although this undoubtedly was necessary at some stage, it is not certain that the old quantum theory need have been in quite such a bad way, and if the new theory had not arisen when it did, a reformulation of the old theory might have carried it along much longer.)

By one of the astonishing coincidences which do occur in science, two such new theories came within a few months of each other in 1925–6. One of these is a fairly straightforward generalisation from what had gone before, although it was a stroke of genius to have hit upon it. Let us describe this first. Another field of physics in which integers tend to occur is that of waves on a stretched string or stretched membrane, and already by 1923 de Broglie (1924) had noticed that in the Bohr–Rutherford model of the hydrogen atom the orbits in which the electron was allowed to move without radiating were exactly those whose length was an integral multiple of the wave length, h/mv, where m is the electron mass and v its velocity. This fact by itself would have been of no significance, but in 1923 Davisson and Kunsman (1923) had reflected a beam of electrons from a platinum plate and showed that, instead of the reflection which one would have expected from a stream of classical particles, a pattern with curious maxima resulted, which was later shown by Elsasser to be consistent with the electrons being scattered like waves of wavelength h/mv.

Instead of considering the rather complicated problem of reflection from a platinum sheet we will consider the simpler experiment in which a stream of electrons passes through two holes in a screen and falls upon another screen which is sensitive to electrons. If such an experiment were carried out it would be found that a series of lines occurred on the screen. The lines occur at the places at which electrons have passed along two paths that are in phase (in the sense of the wavelength just mentioned) and in between the lines are dark spaces corresponding to out-of-phase interference. How can we reconcile such a result with the dynamical description of the electron? Clearly, since the problem here is to understand how a more general dynamical description than usual may be given, we must start with as general as possible a description of dynamics, and show how to alter this. Treating the whole matter non-relativistically, which was the way that it was considered in 1925, we can say that each path through one hole or the other is determined by the equations of motion, which may be written in the form

$$\delta \int L\, dt = 0.$$

Here L is, as usual, the Lagrangian, which, for a free particle, is $T = \frac{1}{2}mv^2$. (The holes are workless constraints, which have to be applied during the calculation of the variation, but which do not enter the Lagrangian.)

The experimental results now show that when we calculate the quantity

$$S = lp$$

(where l is the length and p the momentum) along an actual path there will be bright patches on the screen when the two paths up to the point satisfy $S_1 - S_2 = nh$, and dark patches intermediately. Because this is a rather complicated way of expressing it it is much more convenient to write instead of the quantity which we have just considered the expression

$$\psi = e^{iS/\hbar},$$

where $\hbar = h/2\pi$ is a new constant, which is more convenient than Planck's original one. The condition for brightness is then $\psi_1 = \psi_2$ and for darkness $\psi_1 = -\psi_2$. This in turn may be expressed slightly differently by considering the quantity $\psi = \psi_1 + \psi_2$. This quantity has maximum amplitude, of unity, at a bright point and has its minimum amplitude, which is zero, at a dark point.

Essentially the theory which de Broglie (1926), Schrödinger (1926), Klein (1927) and Gordon (1926) found independently was one which determined the equation satisfied by this function. The analogy here which they exploited is with optics where the first equation of motion is replaced by Fermat's principle of least time but where the correct wave theory is somewhat more complicated. It will be instructive for us to work out a little more of

Schrödinger's theory. The following mathematical development leans somewhat on the last sections of Chapters 2 and 3. The reader who has omitted them is advised to return to them and read them, at least briefly, to give him a résumé of the ideas now presupposed. From the fundamental variational principle of the last chapter we have

$$\delta \int L \, dt = [p_\alpha \, \delta q^\alpha - H \, \delta t] + \int L_\alpha \, \delta q^\alpha \, dt.$$

It will turn out that the integral which is varied on the left-hand side will be what is to be identified with the quantity S above. Writing it as S in anticipation, and considering it *as a function of position in space and time only*, by fixing the starting-point of the integration and then integrating from there to the given point in space and time along a path (so that $L_\alpha = 0$ and the right-hand integral vanishes), it follows that

$$\frac{\partial S}{\partial q^\alpha} = p_\alpha, \quad \frac{\partial S}{\partial t} = -H,$$

since we keep the lower limit fixed. These equations may be used to calculate S for the special case of the free particle for which $\mathbf{p}$ and H are constant, since it gives

$$S = \mathbf{p} \cdot \mathbf{r} - Ht.$$

At any fixed time, then, we have $S = \mathbf{p} \cdot \mathbf{r}$ so that the identification of this S with the one used above in describing the experimental results is fully justified. We now need to convert this in terms of the new function and we notice at once that

$$\frac{\partial \psi}{\partial q^\alpha} = \frac{i}{\hbar} \frac{\partial S}{\partial q^\alpha} \psi, \quad \frac{\partial \psi}{\partial t} = \frac{i}{\hbar} \frac{\partial S}{\partial t} \psi.$$

Consider now the particular case of a particle moving in a potential field in which the Lagrangian has the form

$$L = T - V = \tfrac{1}{2} m v^2 - V.$$

APPLICATIONS IN QUANTUM THEORY 63

(Of course, up to now we have been considering the free particle, $V = 0$, but it involves no more trouble to include a potential field, and the general theory developed so far is independent of any particular form for the Lagrangian.) In such a case we have

$$\mathbf{p} = \frac{\partial L}{\partial \mathbf{v}} = m\mathbf{v},$$

$$H = \mathbf{p} \cdot \mathbf{v} - L = \frac{\mathbf{p}^2}{2m} + V.$$

Substituting from the expressions for momentum and energy in terms of S gives us a certain equation involving the function S, known as the Hamilton–Jacobi equation

$$\frac{1}{2m}(\triangledown S)^2 + \frac{\partial S}{\partial t} + V = 0.$$

Of course, in proceeding from the description in terms of S to the new description it may well be necessary to generalise and alter the theory slightly just as one would have to do in the optical case in order to reach the correct wave description, which would in fact be that of Maxwell's equations. So here, when we substitute for S the function ψ, we must take account of the smallness of Planck's constant in cgs units (which is equivalent to its smallness in terms of ordinary everyday quantities). If we substitute directly, however, we shall be involved in a highly non-linear theory (since the squaring of $\partial \psi / \partial q^\alpha$ will bring in a ψ^2), whereas the wave analogy leads us to expect linear equations, as Maxwell's are. This is evidently a point at which, if we were able to start with the unknown wave theory and proceed in the opposite direction an approximation would have entered. The probable nature of this approximation becomes clear if we calculate a second derivative:

$$\frac{\partial^2 \psi}{\partial q^\alpha \, \partial q^\beta} = \left[\frac{i}{\hbar} \frac{\partial^2 S}{\partial q^\alpha \, \partial q^\beta} + \left(\frac{i}{\hbar}\right)^2 \frac{\partial S}{\partial q^\alpha} \frac{\partial S}{\partial q^\beta} \right] \psi.$$

In the square bracket the first term can be expected to be negligible compared with the second (in view of the smallness of Planck's constant) and so we make the substitutions

$$\nabla^2\psi = \left(\frac{i}{\hbar}\right)^2 (\nabla S)^2 \psi, \quad \frac{\partial \psi}{\partial t} = \frac{i}{\hbar} \frac{\partial S}{\partial t} \psi,$$

in the Hamilton–Jacobi equation, giving

$$\nabla^2\psi + \frac{2mi}{\hbar}\frac{\partial \psi}{\partial t} - \frac{2m}{\hbar^2} V\psi = 0,$$

generally known as Schrödinger's equation. (In the case where the energy is constant, say $H = E$, so that t is only involved in S in the term $-Et$, and so in ψ in the factor $e^{-iEt/\hbar}$, the time-derivative can be removed by defining

$$\psi = u e^{-iEt/\hbar}$$

where u is independent of t, and satisfies

$$\nabla^2 u + \frac{2m}{\hbar^2}(E-V)u = 0,$$

the time-independent Schrödinger equation.)

In June and July of the same year Born (1926) completed the generalisation of the wave theory by giving an interpretation of the function ψ introduced, which is now widely held as giving, by the square of its amplitude, the probability of finding the particle at a particular point. From the point of view of generalising this to special relativity, however, the occurrence of the so-called wave functions and their interpretation is a considerable hindrance, however useful it may be in visualising the physical interpretation. Accordingly we shall turn to the other formulation which was put forward. This alternative theory was suggested by Heisenberg (1925) 8 months earlier than that of wave mechanics (in July) and is much more straightforward to generalise to special relativity. The easiest way of explaining Heisenberg's theory will be to deal with one particular problem by both methods and for this purpose we shall choose the simplest nontrivial one, that is, a harmonic oscillator. The classical equation of motion,

$$m\ddot{x} = -m\omega^2 x,$$

can be derived from the Lagrangian

$$L = \tfrac{1}{2}m(\dot{x}^2 - \omega^2 x^2),$$

or, equivalently, from the Hamiltonian

$$H = \frac{p^2}{2m} + \frac{1}{2}m\omega^2 x^2.$$

The method of dealing with this in the old quantum theory can be slightly generalised by going over to the homogeneous formulation, with t and x as coordinates, for which

$$f\, du = \tfrac{1}{2}m(\dot{x}^2 - \omega^2 x^2)\, dt,$$

i.e.
$$f = \frac{1}{2}m\left(\frac{\dot{x}^2}{\dot{t}}\right) - \omega^2 x^2 \dot{t}.$$

Then the so-called *stationary states* are given by requiring that $\oint f\, du$ round any *closed* path is an integral multiple of h. (This replaces the condition $S_1 - S_2 = nh$, where one path is to be traversed in the opposite direction, so that the former condition might be written $S_1 + (-S_2) = nh$.)

Since, however, as we saw in the last chapter,

$$f = p_a \dot{q}^a$$

this condition can be rewritten

$$\oint p_a\, dq^a = nh$$

round any closed circuit, i.e.

$$\oint (p\, dx - H\, dt) = nh$$

since $p_0 = \partial f/\partial t = -H$. Essentially the only closed circuit is that obtained by following the particle round one complete cycle of its motion, so that

$$2 \int_{-a}^{a} m\omega(a^2-x^2)^{1/2} \, dx = nh.$$

Putting $x = a \sin \theta$, we have

$$4 \int_{0}^{\pi/2} m\omega a^2 \cos^2 \theta \, d\theta = nh,$$

so that the quantum condition given by the old quantum theory is

$$\tfrac{1}{2} m\omega a^2 = n\hbar.$$

Since, however, the total energy E is given by

$$E = \tfrac{1}{2} m\omega^2 a^2,$$

it follows that $E = n\hbar\omega = nh\nu$, where ν is the frequency in cycles per second—so that the rule given above agrees with the one introduced by Planck for the spectrum of radiation.

In the Schrödinger treatment, however, one seeks solutions of the time-independent Schrödinger equation

$$\frac{d^2 u}{dx^2} + \frac{2m}{\hbar^2}\left(E - \frac{1}{2}m\omega^2 x^2\right) u = 0.$$

Such solutions do not exist, at least if we confine our attention to analytic functions which are such that

$$\int_{-\infty}^{\infty} |u|^2 \, dx$$

is finite, except when E has certain special values. For if we begin by making the substitution $u = ve^w$ where v, w are new variables,

APPLICATIONS IN QUANTUM THEORY

the equation becomes

$$v'' + 2w'v' + \left[w'' + (w')^2 + \frac{2m}{\hbar^2}\left(E - \frac{1}{2}m\omega^2 x^2\right)\right]v = 0,$$

and this assumes a more reasonable form if we choose

$$(w')^2 = \left(\frac{m\omega}{\hbar}\right)^2 x^2,$$

i.e. $$w' = \pm \frac{m\omega}{\hbar} x, \quad w'' = \pm \frac{m\omega}{\hbar},$$

so that $$w = \pm \frac{m\omega}{2\hbar} x^2.$$

With the condition that $u \to 0$ at infinity (keeping the convergence of the integral in mind) we take the lower sign, and v has then to satisfy

$$v'' - \frac{2m\omega}{\hbar} xv' + \frac{2m}{\hbar^2}\left(E - \frac{1}{2}\hbar\omega\right)v = 0.$$

It is convenient to make a change of variables, putting $x = (\hbar/m\omega)^{1/2} y$, and the equation becomes

$$v'' - 2yv' + 2\mu v = 0$$

where

$$\mu + \tfrac{1}{2} = E/\hbar\omega.$$

If we seek a polynomial solution,

$$v = \Sigma a_r y^r,$$

of this equation, we find that the coefficients are related by

$$a_{k+2} = -\frac{2(\mu - k)}{(k+1)(k+2)} a_k,$$

and this will indeed be a polynomial solution if the sequence of coefficients terminates, i.e. if μ is an integer, n say. The equation will then have a solution of degree n, and as n is given all integral values from zero upwards, a set of polynomials, say H_0, H_1, $H_2 \ldots$, is derived. Since, however, two polynomials, H_r, H_s say, with different values μ_r, μ_s of μ satisfy

$$\int_{-\infty}^{\infty} e^{-y^2} H_r H_s \, dy \equiv A_{rs} = 0$$

(a fact which may easily be proved by integrating by parts, using the fact that the differential equation implies that

$$2\mu v e^{-y^2} = -(v' e^{-y^2})'$$

it follows that the arbitrary constants can be determined so that

$$A_{rs} = \delta_{rs}.$$

Accordingly any polynomial $\phi(y)$ of degree k can be written in the form

$$\phi(y) = \sum_{s=0}^{k} \lambda_s H_s(y).$$

where, in fact,

$$\lambda_s = \int_{-\infty}^{\infty} e^{-y^2} \phi(y) H_s(y) \, dy,$$

and so the whole set of polynomials H_r form a complete basis to express any analytic function. It follows that there are no *other* values of μ for which there is an analytic solution of the equation, that is, that the possible values of μ are the integers 0, 1, 2, ... The energy levels are therefore of the form $(n+\frac{1}{2})\hbar\omega$, and so the new treatment gives the same answers as the old for the *differences* ($\hbar\omega$) of energy levels, but predicts also a zero-point energy $\frac{1}{2}\hbar\omega$. This new prediction was at first welcomed, as agreeing with some of the spectroscopic evidence, but it has since been a source of acute embarrassment in the development of quantum mechanics.

However that may be, we can now use the wave-mechanical picture to illustrate Heisenberg's approach (though he reached it quite independently, and, indeed, as we have said, earlier). Essentially the problem of solving Schrödinger's equation is not the determination of the function ψ or u; this is merely a technique for determining the values of the constant E for which such a function exists. Abstractly the problem can be expressed thus; consider an operator **H** which is derived from the Hamiltonian by substituting $(\hbar/i)(\partial/\partial q^\alpha)$ for p_α wherever it occurs. Form the equation

$$\mathbf{H}\psi = E\psi,$$

where E is a number; the problem is then the determination of the so-called eigenvalues E. The operator $\mathbf{p}_\alpha$ is connected to the coordinate operator $\mathbf{q}^\beta$ by the *commutation rule* (by substituting $p_\alpha = (\hbar/i)(\partial/\partial q^\alpha)$

$$(\mathbf{p}_\alpha \mathbf{q}^\beta - \mathbf{q}^\beta \mathbf{p}_\alpha)\psi = \frac{\hbar}{i}\delta_\alpha^\beta \psi,$$

and Heisenberg's technique is simply to hold fast to the Hamiltonian operator, and to the commutation rule

$$\mathbf{p}_\alpha \mathbf{q}^\beta - \mathbf{q}^\beta \mathbf{p}_\alpha = \frac{\hbar}{i}\delta_\alpha^\beta,$$

without worrying about the function ψ at all. This much is already clear in Heisenberg's first paper on the subject. However, the later development of the theory by Born, Heisenberg and Jordan required rather complicated manipulations in order to find a suitable form of operator differentiation to put into the equations of motion. Dirac, in the same year (1925), provided a great simplification by noting, firstly, that the only time derivatives entering in the Hamiltonian form of dynamics can be put in Poisson bracket form (as was mentioned in Chapter 3) (since if ϕ is any variable of the motion, then in the notation of Chapter 3,

$$\frac{d\phi}{dt} = \frac{\partial \phi}{\partial X^A}\frac{\partial X^A}{dt} = \frac{\partial \phi}{\partial X^A}\theta^{AB}\frac{\partial H}{\partial X^B} = (\phi, H)$$

and, secondly, that, for large quantum numbers, the operator $\boldsymbol{\phi\psi} - \boldsymbol{\psi\phi}$ formed from the analogues of two classical quantities ϕ, ψ connected with the motion, becomes

$$\boldsymbol{\phi\psi} - \boldsymbol{\psi\phi} = i\hbar(\phi, \psi).$$

Accordingly Dirac *defined*

$$\frac{d\boldsymbol{\phi}}{dt} = \frac{i}{\hbar}(\mathbf{H}\boldsymbol{\phi} - \boldsymbol{\phi}\mathbf{H}),$$

and so in the case of the harmonic oscillator, where

$$\mathbf{H} = \mathbf{p}^2/2m + \tfrac{1}{2}m\omega^2\mathbf{q}^2,$$

it follows that

$$\begin{aligned}
\mathbf{Hq} - \mathbf{qH} &= \frac{1}{2m}(\mathbf{p}^2\mathbf{q} - \mathbf{qp}^2) \\
&= \frac{1}{2m}[\mathbf{p}(\mathbf{pq} - \mathbf{qp}) + (\mathbf{pq} - \mathbf{qp})\mathbf{p}] \\
&= \frac{\hbar}{mi}\mathbf{p}, \\
\mathbf{Hp} - \mathbf{pH} &= \tfrac{1}{2}m\omega^2[\mathbf{q}(\mathbf{qp} - \mathbf{pq}) + (\mathbf{qp} - \mathbf{pq})\mathbf{q}] \\
&= i\hbar m\omega^2 \mathbf{q}
\end{aligned}$$

in complete analogy with the classical equations of motion.

Suppose now that the oscillator is in a stationary state; the Hamiltonian operator can be looked on as an infinite matrix operator acting on the components of a function referred to the basis vectors $e^{-y^2/2}H_r(y)$ mentioned above. And if these are the states corresponding to the eigenvalues, the matrix will be diagonal. As a result the above equations become, in matrix notation,

$$\Sigma(H_{rs}q_{st} - q_{rs}H_{st}) = \frac{\hbar}{mi}p_{rt},$$

or, if **H** is represented by the diagonal matrix

$$\mathbf{H} = \begin{pmatrix} \lambda_1 & 0 & 0 & \cdots \\ 0 & \lambda_2 & 0 & \cdots \\ 0 & 0 & \lambda_3 & \cdots \\ \cdot & \cdot & \cdot & \cdot \\ \cdot & \cdot & \cdot & \cdot \\ \cdot & \cdot & \cdot & \cdot \end{pmatrix}$$

then
$$(\lambda_r - \lambda_t) q_{rt} = \frac{\hbar}{mi} p_{rt},$$

and
$$(\lambda_r - \lambda_t) p_{rt} = i\hbar m\omega^2 q_{rt},$$

which give
$$[(\lambda_r - \lambda_t)^2 - (\hbar\omega)^2] q_{rt} = 0.$$

Thus any non-zero elements of **q** must correspond to eigenvalues λ differing by $\hbar\omega$; so that, if the eigenvalues are arranged in ascending order of magnitude, they are $\alpha, \alpha+\hbar\omega, \alpha+2\hbar\omega, \ldots$ and the only non-zero values of q_{rs}, p_{rs} are where r, s differ by unity. It only remains to find the value of α, which is H_{11}, the lowest energy level. Now, substituting,

$$q_{12} = \frac{i}{m\omega} p_{12} \quad \text{and} \quad q_{21} = -\frac{i}{\omega m} p_{21},$$

which give, in
$$\mathbf{pq} - \mathbf{qp} = -i\hbar,$$

that
$$q_{12} q_{21} = \frac{\hbar}{2m\omega},$$

and therefore also $p_{12} p_{21} = \frac{1}{2}\hbar m\omega$. Since, however,

$$H_{11} = \frac{1}{2m} p_{12} p_{21} + \frac{1}{2} m\omega^2 q_{12} q_{21},$$

we find that $H_{11} = \alpha = \frac{1}{2}\hbar\omega$. There is thus complete agreement between the two approaches.

It is now time to consider how all this may be related to the Lorentz transformation. Neither the original formulation of

Schrödinger nor that of Heisenberg had produced a relativistically invariant theory. Dirac, looking at the problem more from the Schrödinger angle, noticed that for a free particle Schrödinger's equation was of the first order in the time. Since the purpose of the equation was to show the time evolution of the system, this was a necessity. On the other hand, it was of the second order in the space variables, and this meant that it was hopeless to try to render it relativistically invariant. Dirac therefore set about formulating an equation whose solutions would be as much as possible like those of the Schrödinger equation but which would be of the first order in the space and time variables. His success in this respect can be seen from his paper, reproduced as Extract 8.

In our notation, what Dirac does there is to consider, instead of the Schrödinger equation which is the operator form of the non-relativistic energy equation

$$\mathbf{p}^2/2m + V = E,$$

the so-called Klein–Gordon equation which is the operator form of the relativistic expression for the *total* energy (including the rest-energy)

$$E = m'c^2$$
$$= mc^2 \left(\frac{1}{\sqrt{(1-v^2/c^2)}} \right)$$
$$= [\mathbf{p}^2 c^2 + m^2 c^4]^{1/2},$$

that is,

$$E^2 - \mathbf{p}^2 c^2 = m^2 c^4.$$

The Klein–Gordon equation is then

$$\left[\left(\frac{\hbar}{ic} \frac{\partial}{\partial t} \right)^2 - \left(\frac{\hbar}{i} \nabla \right)^2 \right] \psi = m^2 c^2 \psi$$

or

$$\left(\Box^2 + \frac{m^2 c^2}{\hbar^2} \right) \psi = 0,$$

APPLICATIONS IN QUANTUM THEORY 73

writing $\Box^2$ for the usual wave-equation operator:

$$\Box^2 = \frac{1}{c^2}\frac{\partial^2}{\partial t^2} - \nabla^2.$$

This is now Lorentz invariant, but has the new disadvantage of being of the second order in the time. Dirac assumed, writing the equation in the form

$$\eta^{\alpha\beta}\frac{\partial}{\partial x^\alpha}\frac{\partial}{\partial x^\beta}\psi + \frac{m^2c^2}{\hbar^2}\psi = 0,$$

that there exist symbols γ^μ which will factorise the operator, i.e. such that

$$\eta^{\alpha\beta}\frac{\partial^2}{\partial x^\alpha \partial x^\beta} = \left(\gamma^\mu \frac{\partial}{\partial x^\mu}\right)\left(\gamma^\nu \frac{\partial}{\partial x^\nu}\right).$$

If the γ^μ are constants, they must then satisfy

$$\gamma^\mu\gamma^\nu + \gamma^\nu\gamma^\mu = 2\eta^{\mu\nu}.$$

It is possible to satisfy such a set of conditions by means of 4×4 matrices, as will be clear later in the chapter. Accordingly the wave function ψ will now have to have four components ψ_r (say), so that

$$\gamma^\mu \frac{\partial}{\partial x^\mu}\psi$$

means the set

$$\left\{\sum_s \gamma^\mu_{rs}\frac{\partial \psi_s}{\partial x^\mu}\right\} \quad \text{for } r = 1, \ldots 4.$$

Now not only did Dirac's reformulation achieve the desired Lorentz invariance. It led also to two important experimental predictions. The first of these is fairly easy to explain. The equation derived by Dirac is, as we have seen, got from a quadratic expression for the energy

$$E^2 = p^2c^2 + m^2c^4,$$

by factorising it by means of the introduction of the new quantities. When a square root is taken in this way there is an ambiguous sign to be determined. Dirac's original idea was that the solutions of positive energy were to be taken, and the ones of negative energy to be disregarded. However, Klein (1929) showed that in quite a simple experimental situation it was possible for particles to jump from negative to positive energy states or vice versa. Essentially this is because quantum mechanics predicts that a potential barrier, which would repel all particles of a particular energy in the classical theory, no longer has this property in the quantum case ("tunnel effect"). Instead there is a small but definite probability that the particle will penetrate the barrier. Such a barrier could be one keeping apart positive and negative energy states. In such a situation it is impossible to disregard the negative energy states, and Dirac was forced to the "hole theory" in which the vacuum was regarded as the state of affairs in which every negative energy state contains one electron and all the positive energy states are empty (fortunately, since the particles concerned are electrons, the Pauli exclusion principle forbids there being more than one electron in any state, so this is a meaningful prescription). When an electron jumps from a negative energy state to a positive one a "hole" is left behind, that is to say, a state of affairs in which the vacuum has less negative charge than it should do. In other words, the hole theory predicts the existence of another particle with charge equal but opposite to that of the electron and of the same mass. This is a *positron*, which was discovered experimentally by Anderson (1932) (see Extract 9 of the present book), though it is noteworthy that Anderson was working in complete ignorance of Dirac's theoretical prediction and was indeed trying to interpret the cloud chamber tracks as those of protons.

The other experimental confirmation requires rather more explanation. Indeed the form in which this prediction was originally found in Dirac's 1928 paper (Extract 8 of the present volume) is somewhat unsatisfactory. The following method of derivation

(Rund, 1966) is intended to rectify and generalise Dirac's deduction. Firstly, as can be seen above, Dirac begins by making the assumption that the Hamiltonian operator has a linear form in the generalised momenta. That is to say he makes the assumption:

$$\mathcal{H} = \Gamma^a \pi_a,$$

where Γ^b commutes with π_a.

It will clarify the treatment of the Dirac theory if we forget all about the differential operator form for the momenta, since it plays no important part in the calculations. Instead we simply consider any Hamiltonian which is *linear* in the momenta; this is evidently very different in one respect from a classical system since, reverting to the end of Chapter 3, where

$$\mathcal{H} = \tfrac{1}{2} \pi_a \dot{q}^a$$

it seems to follow here that $\Gamma^a = \dot{q}^a$, and, since the Γ^a are constants, this corresponds to motion with constant velocity. However, the Γ^a no longer obey the commutative law of multiplication, so that this inference is no longer valid, and indeed the whole connection of the Hamiltonian and Lagrangian formalisms is altered. For this reason we can no longer appeal to Noether's theorem in our search for conserved quantities; and the investigation that follows can be regarded as the nearest substitute for Noether's theorem, assuming *only* the operator form for the Hamiltonian.

In this formulation, then, the Hamiltonian is an operator, and in order to discuss transformations under the Lorentz group it must operate on some function ψ which transforms under a representation of the group. Further, $\mathcal{H}\psi$ must be well defined and must transform under the same representation as ψ. These two restrictions determine something about the algebra of the expressions Γ^a, which are assumed in Dirac's theory to be invariant quantities. Just what is determined can be found as follows.

Consider an infinitesimal coordinate transformation

$$x^a \rightarrow x^{a'} = (\delta_a^{a'} + \varepsilon_a^{a'}) x^a;$$

the wave functions must then transform under some transformation very near to the identity, say of the form $\psi' = \mathbf{\Lambda}\psi$, where $\mathbf{\Lambda} = \mathbf{I} + \frac{1}{2}\varepsilon_{ab}\mathbf{I}^{ab}$. Of course, in conformity with the transformation of the coordinates, the generalised momenta will transform under:

$$\pi_{a'} = (\delta^a_{a'} + \varepsilon^a_{a'})\pi_a.$$

Consider now the transformation of $\phi = \mathscr{H}\psi$. On the one hand, we know that:

$$\phi' = \mathbf{\Lambda}\phi = (\mathbf{I} + \tfrac{1}{2}\varepsilon_{cd}\mathbf{I}^{cd})\phi;$$

on the other hand,

$$\mathbf{\Lambda}\phi = \mathscr{H}'\psi' = \Gamma^{a'}(\delta^a_{a'} + \varepsilon^a_{a'})\pi_a\mathbf{\Lambda}\psi = \mathbf{\Lambda}\Gamma^a\pi_a\psi$$

and this gives as a result:

$$\Gamma^{a'}(\delta^a_{a'} + \varepsilon^a_{a'})\mathbf{\Lambda} = \mathbf{\Lambda}\Gamma^a.$$

Substituting:

$$\Gamma^{a'}(\delta^a_{a'} + \varepsilon^a_{a'})(\mathbf{I} + \tfrac{1}{2}\varepsilon_{cd}\mathbf{I}^{cd}) = (\mathbf{I} + \tfrac{1}{2}\varepsilon_{cd}\mathbf{I}^{cd})\Gamma^a.$$

That is to say,

$$\Gamma^{a'}\varepsilon^a_{a'} + \tfrac{1}{2}\varepsilon_{cd}\Gamma^{a'}\mathbf{I}^{cd} = \tfrac{1}{2}\varepsilon_{cd}\mathbf{I}^{cd}\Gamma^a,$$

i.e.

$$\tfrac{1}{2}\varepsilon_{cd}(\mathbf{I}^{cd}\Gamma^a - \Gamma^a\mathbf{I}^{cd}) = \varepsilon_{cd}\Gamma^{a'}\delta^c_a\eta^{da}$$
$$= \tfrac{1}{2}\varepsilon_{cd}(\eta^{da}\Gamma^c - \eta^{ca}\Gamma^d).$$

Thus the requirements of invariance under the Lorentz group lead to the relations:

$$\mathbf{I}^{cd}\Gamma^a - \Gamma^a\mathbf{I}^{cd} = \eta^{da}\Gamma^c - \eta^{ca}\Gamma^d,$$

between the original quantities introduced by Dirac and the new quantities $\mathbf{I}^{cd}$. It remains to understand what these quantities $\mathbf{I}^{cd}$ are.

For this purpose consider the expression for the analogue of the angular momentum (*analogue*, because the expression concerned has six components, three of which correspond to the usual angu-

APPLICATIONS IN QUANTUM THEORY 77

lar momentum, and three which do not). When we calculate the rate of change of this quantity in accordance with the rules given for calculating derivatives by Dirac, we get

$$\frac{d\mathbf{M}^{\alpha\beta}}{ds} = [\mathbf{M}^{\alpha\beta}, \mathcal{H}]$$

$$= \frac{i}{\hbar} \{[\mathbf{x}^{\alpha}\eta^{\beta\mu} - \mathbf{x}^{\beta}\eta^{\alpha\mu}]\boldsymbol{\pi}_{\mu}\Gamma^{a}\boldsymbol{\pi}_{a} - \Gamma^{a}\boldsymbol{\pi}_{a}(\mathbf{x}^{\alpha}\eta^{\beta\mu} - \mathbf{x}^{\beta}\eta^{\alpha\mu})\boldsymbol{\pi}_{\mu}\}$$

$$= \frac{i}{\hbar} \Gamma^{a} \{\eta^{\beta\mu}(\mathbf{x}^{\alpha}\boldsymbol{\pi}_{\mu}\boldsymbol{\pi}_{a} - \boldsymbol{\pi}_{a}\mathbf{x}^{\alpha}\boldsymbol{\pi}_{\mu}) - \eta^{\alpha\mu}(\mathbf{x}^{\beta}\boldsymbol{\pi}_{\mu}\boldsymbol{\pi}_{a} - \boldsymbol{\pi}_{a}\mathbf{x}^{\beta}\boldsymbol{\pi}_{\mu})\}$$

$$= \frac{i}{\hbar} \Gamma^{a} \{\eta^{\beta\mu}[\mathbf{x}^{\alpha}, \boldsymbol{\pi}_{a}]\boldsymbol{\pi}_{\mu} - \eta^{\alpha\mu}[\mathbf{x}^{\beta}, \boldsymbol{\pi}_{a}]\boldsymbol{\pi}_{\mu}\}$$

$$= \frac{i}{\hbar} \Gamma^{a} \{\boldsymbol{\pi}^{\beta}[\mathbf{x}^{\alpha}, \boldsymbol{\pi}_{a}] - \boldsymbol{\pi}^{\alpha}[\mathbf{x}^{\beta}, \boldsymbol{\pi}_{a}]\}.$$

At first sight this is surprising, since the rate of change of angular momentum has ceased to be zero. The expression can, however, be simplified by using the usual commutation rules

$$[\mathbf{x}^{\alpha}, \boldsymbol{\pi}_{a}] = i\hbar \delta^{\alpha}_{a},$$

so that in all:

$$\frac{d\mathbf{M}^{\alpha\beta}}{ds} = -(\Gamma^{\alpha}\boldsymbol{\pi}^{\beta} - \boldsymbol{\pi}^{\alpha}\Gamma^{\beta}).$$

Recalling now the quantities $\mathbf{I}^{cd}$, introduced by the transformation of the wave function, it follows from above that

$$\mathbf{I}^{cd}\mathcal{H} - \mathcal{H}\mathbf{I}^{cd} = \boldsymbol{\pi}^{d}\Gamma^{c} - \boldsymbol{\pi}^{c}\Gamma^{d},$$

so that

$$\frac{d\mathbf{M}^{\alpha\beta}}{ds} = -(\mathbf{I}^{\alpha\beta}\mathcal{H} - \mathcal{H}\mathbf{I}^{\alpha\beta}) = -\frac{d\mathbf{I}^{\alpha\beta}}{ds}.$$

The quantity $\mathbf{I}^{cd}$ is known in quantum theory as the *spin angular momentum*, and is regarded as an intrinsic property of the particle. The equation derived then shows that the total angular momentum, that is the sum of the spin angular momentum and the ordinary angular momentum, produced by the momentum in virtue of the position of the particle relative to the origin, is the quantity which is conserved.

It is important to notice that the spin angular momentum depends on the transformation properties of the wave function. For example, if the wave function is a scalar the infinitesimal transformations of it are all identity transformations. The quantities $\mathbf{I}^{cd}$ are zero and the spin angular momentum accordingly vanishes. The particles defined by scalar wave functions therefore have no spin. Again, if the wave function is a vector, for example, the quantities $\mathbf{I}^{cd}$ can be found as follows: the transformation

$$\psi_{a'} = (\delta^a_{a'} + \varepsilon^a_{a'})\psi_a$$

is to be rewritten in the form

$$\psi' = (\mathbf{I} + \tfrac{1}{2}\varepsilon_{rs}\mathbf{I}^{rs})\psi,$$

so that

$$\delta\psi_a = \psi_{a'} - \psi_a = \varepsilon_{ab}\eta^{bc}\psi_c = \tfrac{1}{2}\varepsilon_{rs}(\mathbf{I}^{rs})^c_a\psi_c.$$

Hence

$$\tfrac{1}{2}\varepsilon_{rs}(\mathbf{I}^{rs})^c_a = \tfrac{1}{2}\varepsilon_{rs}(\eta^{sc}\delta^r_a - \eta^{rc}\delta^s_a),$$

i.e.

$$(\mathbf{I}^{rs})^c_a = \delta^r_a \eta^{sc} - \delta^s_a \eta^{rc}.$$

For example, I^{12} has the matrix form

$$I^{12} = \begin{bmatrix} \cdot & \cdot & \cdot & \cdot \\ \cdot & \cdot & 1 & \cdot \\ \cdot & -1 & \cdot & \cdot \\ \cdot & \cdot & \cdot & \cdot \end{bmatrix}.$$

The corresponding spin angular momentum $\mathbf{I}^{rs}$ is the one appropriate to particles having vector wave functions.

APPLICATIONS IN QUANTUM THEORY 79

As well as the tensor transformations, however, quantities transforming under representations of a different kind arise. These quantities already exist in a two-dimensional theory, and it will be instructive to consider them here. Since the rotation group in three dimensions is a subgroup of the Lorentz group, and the rotation group in two dimensions is a subgroup of the rotation group in three, the quantities introduced in this section will be equally important in a study of the Lorentz group. Consider then a rotation in two dimensions of the form

$$x^{1'} = x^1 \cos\theta + x^2 \sin\theta,$$
$$x^{2'} = -x^1 \sin\theta + x^2 \cos\theta.$$

The vector representation of this rotation has the form $A^{\alpha'} = l_\alpha^{\alpha'} A^\alpha$, where the components can be read off from the fact that the componets of the vector transform in just the same way as the coordinates. Similarly the tensor of rank 2 has the transformation law $A^{\alpha'\beta'} = l_\alpha^{\alpha'} l_\beta^{\beta'} A^{\alpha\beta}$ which, written out at length, is

$$A^{1'1'} = A^{11} \cos^2\theta + (A^{12} + A^{21}) \cos\theta \sin\theta + A^{22} \sin^2\theta,$$
$$A^{1'2'} = A^{12} \cos^2\theta - A^{21} \sin^2\theta + (A^{22} - A^{11}) \cos\theta \sin\theta,$$
$$A^{2'1'} = A^{21} \cos^2\theta - A^{12} \sin^2\theta + (A^{22} - A^{11}) \cos\theta \sin\theta,$$
$$A^{2'2'} = A^{22} \cos^2\theta - (A^{12} + A^{21}) \cos\theta \sin\theta + A^{11} \sin^2\theta.$$

A glance at these equations shows at once the existence of an invariant quantity, $A^{1'2'} - A^{2'1'} = A^{12} - A^{21}$. Since any tensor can be written as the sum of a symmetric and an anti-symmetric part, by the formula

$$A^{\alpha\beta} = \tfrac{1}{2}(A^{\alpha\beta} + A^{\beta\alpha}) + \tfrac{1}{2}(A^{\alpha\beta} - A^{\beta\alpha}),$$

we can use the invariance in two dimensions of the antisymmetric part to confine our attention to symmetric tensors only. Consider then a *symmetric* tensor and introduce the abbreviated notation

$$A^{11} = p \quad A^{12} = A^{21} = \tfrac{1}{2}q \quad A^{22} = r$$

80 SPECIAL RELATIVITY

so that the transformations become:

$$p' = p \cos^2 \theta + r \sin^2 \theta + \tfrac{1}{2} q \sin 2\theta,$$
$$r' = r \cos^2 \theta + p \sin^2 \theta - \tfrac{1}{2} q \sin 2\theta.$$

It now appears that there is another invariant

$$p' + r' = p + r,$$
and
$$q' = q (\cos^2 \theta - \sin^2 \theta) + (r - p) \sin 2\theta,$$

so that apart from this invariant the remaining quantities transform under the equations

$$q' = q \cos 2\theta + (r - p) \sin 2\theta,$$
$$r' - p' = (r - p) \cos 2\theta - q \sin 2\theta.$$

What is most remarkable about these equations is their strong similarity to the original equations of rotation for vectors, the difference being only that, where the original equations involved functions of a given angle, the new ones involve functions of double the angle. We can look at all this from a different, and highly instructive, point of view if we imagine the *new* equations to be the ones defining the rotation of a vector, and so replace 2θ by the angle of rotation, which may again be called θ. The argument above then shows that such a vector is derived from *other* quantities, ϕ^1, ϕ^2 say, transforming under by the rule

$$\phi^{1\prime} = \phi^1 \cos \theta/2 + \phi^2 \sin \theta/2,$$
$$\phi^{2\prime} = -\phi^1 \sin \theta/2 + \phi^2 \cos \theta/2.$$

Clearly this rule, from the way in which it has been derived, must correspond to a representation of the rotation group. But it is a representation of a different form from the tensor transformations, because the two rotations defined by θ and $\theta + 2\pi$ are identical, as members of the rotation group, but give rise to different objects in the representation, differing in fact by sign. It is therefore an example of a *two-valued representation*.

When the corresponding theory is extended to the whole of the Lorentz group instead of the three-dimensional orthogonal group it is necessary either to have a pair of such objects or, alternatively, to have a single object with four components. These objects (whether of two or four components) are each called *spinors*. Dirac chose the spin representation for the wave function of the electron, although not because it was obligatory to do so in terms of his formalism, as seemed to be the case at the time. It is clear by looking at the infinitesimal operators:

$$x^{1\prime} = x^1 + \theta x^2,$$
$$x^{2\prime} = -\theta x^1 + x^2,$$
$$\psi^{1\prime} = \psi^1 + \tfrac{1}{2}\theta \psi^2,$$
$$\psi^{2\prime} = \psi^2 - \tfrac{1}{2}\theta \psi^1,$$

and comparing with the calculation for vector wave-functions, that the spin angular momentum operators for spinor wave-functions are *half* of those for vector wave functions. In quantum mechanics the spin is measured, for historical reasons, in units determined by the vector wave-functions, so that for spinor particles the spin is one-half. Thus the choice of a two-valued representation by Dirac for the wave function was in fact an experimental result; the "observed" electron spin (in a classical analogue)—observed, that is, by its effect on spectral lines—was only one-half of what would be predicted for vector wave-functions. This discussion of conservation laws may be regarded as the appropriate generalisation of Noether's theorem (see Chapter 2) to the case in which the Hamiltonian is generalised to contain matrices.

It is clear that further progress in understanding this part of the theory requires a very careful consideration of the representations of the Lorentz group. So far we have been considering finite-dimensional representations. It is also possible to look at the infinite dimensional representations. The whole work was carried out in a definitive fashion by Wigner (1939) and parts of his work form

Extract 10 of the present volume. It would take us too far to try in the space available to us to summarise all the corresponding consequences of special relativity for quantum mechanics, and instead we will content ourselves by showing how the study which we have made of two-valued representations leads us in a natural way to a matrix form of the quantities γ_μ introduced by Dirac.

Let us return to the way in which a vector was used (rotating through an angle 2θ) to lead to the spinor quantities. The components of the vector are given in terms of a pair of quantities like this by the equations

$$V^1 = q = \phi^1\psi^2 + \phi^2\psi^1,$$
$$V^2 = r-p = \phi^2\psi^2 - \phi^1\psi^1,$$

or in terms of new quantities defined below as

$$V^\alpha = \sigma^\alpha_{\beta\gamma}\phi^\beta\psi^\gamma.$$

The quantities $\sigma^\alpha_{\beta\gamma}$ introduced in this equation are:

$$\sigma^1_{21} = \sigma^1_{12} = 1, \quad \sigma^2_{22} = 1, \quad \sigma^2_{11} = -1,$$

and these can conveniently be written as two matrices:

$$\sigma^1 = \begin{bmatrix} . & 1 \\ 1 & . \end{bmatrix} \qquad \sigma^2 = \begin{bmatrix} -1 & . \\ . & 1 \end{bmatrix}.$$

At the same time it may be noted that the second of the invariants found above has the form

$$\varepsilon_{\alpha\beta}\phi^\alpha\psi^\beta = \text{invariant},$$

where
$$\varepsilon_{\alpha\beta} = \begin{bmatrix} . & 1 \\ -1 & . \end{bmatrix}.$$

The matrices now introduced have the properties that:

$$(\sigma^1)^2 = (\sigma^2)^2 = 1$$

$$\sigma^1\sigma^2 = \begin{bmatrix} . & 1 \\ -1 & . \end{bmatrix} = -\sigma^2\sigma^1 = -\begin{bmatrix} . & -1 \\ 1 & . \end{bmatrix} = \sigma^3 = \varepsilon$$

$$(\sigma^3)^2 = -\sigma^1\sigma^2\sigma^2\sigma^1 = -1, \quad \sigma^3_{\alpha\beta} = \varepsilon_{\alpha\beta}.$$

APPLICATIONS IN QUANTUM THEORY

There is a certain lack of symmetry in these definitions, and it is therefore appropriate to take instead the three quantities

$$\mathbf{e}_1, \quad \mathbf{e}_2, \quad \mathbf{e}_3 = i\sigma^1, \quad i\sigma^2, \quad \sigma^3.$$

Since we have constructed this whole theory by considering those particular members of the rotation group which leave invariant the z-axis, the quantities which arise have the z-axis as a preferred direction. However, Euler's theorem shows that any member of the rotation group does consist of a rotation about a certain axis so that this specialisation must be one only in appearance, and the theory which we have devised must be adequate for the whole rotation group. This is made clear when we use the quantities just defined to represent a vector in three dimensions in the form

$$\mathbf{a} = a^c \mathbf{e}_c, \quad \text{where now} \quad \mathbf{e}_a \mathbf{e}_b = -\delta_{ab} + \varepsilon_{abc} \mathbf{e}_c.$$

The "product" of two vectors could then be written in terms of the usual scalar product $\mathbf{a} \cdot \mathbf{b}$ and the vector product $\mathbf{a} \wedge \mathbf{b}$, as

$$\mathbf{ab} = -\mathbf{a} \cdot \mathbf{b} + \mathbf{a} \wedge \mathbf{b}.$$

The transforms of the rotation group leave the lengths of vectors and the angles between them unaltered, and therefore must preserve such products. In other words, the transformations of the form

$$\mathbf{a} \rightarrow \mathbf{a}' = q\mathbf{a}q^{-1},$$

where q has the form,

$$q = \cos \varphi + \varepsilon_3 \sin \phi,$$

must be all members of the rotation group, and since the group of all transformations so defined is in fact a three parameter group, one may well expect it to be the whole of the rotation group. The fact that this is so can be seen as follows: the choice of $\mathbf{e}_3$ as the unit vector in q is evidently not of any great significance, since the z-axis can always be chosen in an appropriate direction. If, then,

$$q = \cos \phi + \mathbf{e}_3 \sin \phi,$$

it is clear by multiplying out that

$$q^{-1} = \cos\phi - \mathbf{e}_3 \sin\phi,$$

so that the formula is

$$\begin{aligned}
\mathbf{a}' &= (\cos\phi + \mathbf{e}_3 \sin\phi)(a_1\mathbf{e}_1 + a_2\mathbf{e}_2 + a_3\mathbf{e}_3)(\cos\phi - \mathbf{e}_3 \sin\phi) \\
&= (\cos\phi + \mathbf{e}_3 \sin\phi)^2 (a_1\mathbf{e}_1 + a_2\mathbf{e}_2) + a_3\mathbf{e}_3 \\
&= (\cos 2\phi + \mathbf{e}_3 \sin 2\phi)(a_1\mathbf{e}_1 + a_2\mathbf{e}_2) + a_3\mathbf{e}_3 \\
&= (a_1 \cos 2\phi - a_2 \sin 2\phi)\mathbf{e}_1 + (a_1 \sin 2\phi + a_2 \cos 2\phi)\mathbf{e}_2 + a_3\mathbf{e}_3
\end{aligned}$$

which is a general rotation about the z-axis.

The quantities $\mathbf{e}_1$, $\mathbf{e}_2$, $\mathbf{e}_3$ (or equivalently σ_1, σ_2, σ_3) behave to some extent like the γ^μ introduced by Dirac. For example, the commutation rule

$$\mathbf{e}_a \mathbf{e}_b - \mathbf{e}_b \mathbf{e}_a = -2\delta_{ab}$$

may be compared with

$$\gamma^\mu \gamma^\nu + \gamma^\nu \gamma^\mu = 2\eta^{\mu\nu}.$$

The principal difference is the increase in numbers from 3 to 4. However, there is a well-known algebraic trick for increasing the number of elements in a linear algebra, i.e. taking the direct product. That is to say, if a, b, a', b' all belong to some set in which associative multiplication is defined, one can construct a new set by taking pairs of the old ones, as (a, a'), (b, b'), and defining the new set to have the associative multiplication

$$(a, a')(b, b') = (ab, a'b'),$$

whilst multiplication by scalar quantities is defined by

$$(\lambda a, \mu b) = \lambda\mu(a, b)$$

(the fact that these rules are again associative being obvious). With the help of this construction, if

$$K_{11} = (\mathbf{e}_1, \mathbf{e}_1), \quad K_{21} = (\mathbf{e}_2, \mathbf{e}_1), \quad K_{31} = (\mathbf{e}_3, \mathbf{e}_1),$$

APPLICATIONS IN QUANTUM THEORY 85

then clearly K_{11}, K_{21}, K_{31} anticommute and

$$K_{11}^2 = K_{21}^2 = K_{31}^2 = 1$$

(where 1 now denotes the ordered pair (1, 1)). A fourth member anticommuting with each of these three would then be, for instance, $K_{02} = (1, \mathbf{e}_2)$, and $K_{02}^2 = -1$. It is therefore possible to represent the γ^μ by

$$\gamma^1 = iK_{11}, \quad \gamma^2 = iK_{21}, \quad \gamma^3 = iK_{31}, \quad \gamma^4 = iK_{02}.$$

It only remains to show how this leads to a matrix representation of the γ^μ. For this purpose let us define some new matrices, e_{rs}, where e_{rs} is the matrix whose element in the rth row and sth column is 1 and all other of whose elements are zero. These matrices satisfy the obvious multiplication table

$$e_{rs}e_{tu} = \delta_{st}e_{ru}.$$

Now each matrix form of the $\mathbf{e}_a$ can be written

$$\mathbf{e}_a = \Sigma c_{rs} e_{rs}$$

where only two of the c_{rs} are non-zero, and those are ± 1 or $\pm i$. Similarly the matrix form of 1 in 2×2 matrices is $(e_{11}+e_{22})$. In order to apply this to the direct product, consider two sets e_{rs}, f_{rs} to serve as bases of the first and second members of the pairs. Then

$$(e_{rs}f_{tu})(e_{ab}f_{cd}) = \delta_{sa}\delta_{uc}e_{rb}f_{td}.$$

But we could renumber the quantities $e_{rs}f_{tu}$ by defining, first, $E_{rtsu} = e_{rs}f_{tu}$ (note the reversal of s and t on the left-hand side), and then renumbering the pairs (r, t) in an arbitrary way, e.g. by

$$(1, 1) \to 1, \quad (1, 2) \to 2, \quad (2, 1) \to 3, \quad (2, 2) \to 4,$$

and $$(r, t) \to \alpha,$$
so that $$E_{rtsu} \to E_{\alpha\beta}$$
where $$E_{\alpha\beta}E_{\gamma\delta} = \delta_{\beta\gamma}E_{\alpha\delta}$$

according to the formula above.

It is now a simple matter to work out matrix forms for the γ^μ. For example, since
$$\mathbf{e}_1 = i(e_{21}+e_{12}),$$
it follows that
$$\begin{aligned}K_{11} &= -(e_{21}+e_{12})(f_{21}+f_{12})\\ &= -(e_{21}f_{21}+e_{12}f_{21}+e_{21}f_{12}+e_{12}f_{12})\\ &= -(E_{2211}+E_{1221}+E_{2112}+E_{1122})\\ &= -(E_{41}+E_{23}+E_{32}+E_{14}),\end{aligned}$$
and so
$$\gamma^1 = \begin{bmatrix} \cdot & \cdot & \cdot & -i \\ \cdot & \cdot & -i & \cdot \\ \cdot & -i & \cdot & \cdot \\ -i & \cdot & \cdot & \cdot \end{bmatrix}.$$

Similarly for the other γ^μ; this process shows that a 4×4 matrix representation is possible and exhibits one such. But there is, of course, considerable arbitrariness in it, and many other such representations can be found. For particular purposes some of these may be more convenient than others, although they are mathematically equivalent.

References

ANDERSON, C. D. (1932) *Phys. Rev.* **41**, (2) 405 (Extract 9 of the present book).
BOHR, N. (1913) *Phil. Mag.* **26**, 1.
BORN, M. (1926) *Zeit. für Phys.* **38**, 803.
BRADLEY, J. (1728) *Phil. Trans.* **35**, 637.
DE BROGLIE, L. (1924) Doctoral thesis. Reprinted as *Ann. de Phys.* **3**, 22 (1925).
DE BROGLIE, L. (1926) *Jour. de Phys.* **7**, 331.
DAVISSON, C. J. and KUNSMAN, C. H. (1923) *Phys. Rev.* **22**, 242.
DESCARTES, R. (1637) *Dioptrique*, Leyden. (W. Snell died in 1626 without publishing his law, though some think Descartes saw Snell's manuscript.)
DIRAC, P. A. M. (1925) *Proc. Roy. Soc.* (A), **109**, 642.
DIRAC, P. A. M. (1928) *Proc. Roy. Soc.* (A) **117**, 610 (Extract 8 of the present book).
EINSTEIN, A. (1905) *Ann. der Phys.* **17**, 891 (Extract 5 of the present book).
EINSTEIN, A. (1916) *Ann. der Phys.* **49**, 769.
EULER, L. (1750) St. Petersburg Academy.
FERMAT, P. DE (1657) in a letter to Cureau de la Chambre (see *Oeuvres de Fermat*, Paris, 1891).
FIZEAU, H. L. (1859) *Ann. de Chimie*, **57**, 385.
GALILEO, G. (1632) *Dialogue concerning two chief world systems*, Florence.
GORDON, W. (1926) *Zeit. für Phys.* **40**, 117.
TER HAAR, D. (1967) *The Old Quantum Theory*, Pergamon, Oxford.
HAMILTON, W. R. (1835) *Phil. Trans.*, p. 247.
HEISENBERG, W. (1925) *Zeit. für Phys.* **33**, 879. (The full details of the calculations, as given in the text, lean heavily also on BORN, M. and JORDAN, P., *Zeit. für Phys.* **34**, 858 (1925).)
HESSE, M. B. (1955) *Isis* **46**, 337.
KAUFMANN, W. (1902) *Göttinger Nachrichten*, p. 143.
KLEIN, O. (1927) *Zeit. für Phys.* **41**, 407.
KLEIN, O. (1929) *Zeit. für Phys.* **53**, 157.
LAGRANGE, J. L. (1788) *Mécanique analytique*, Paris. (A recent edition incorporating notes by Bertrand and Darboux appeared 1965, Blanchard: Paris.)
LANDSBERG, P. T. (1961) *Nature, London* **189**, 654.
LARMOR, J. (1900) *Aether and Matter*, Cambridge University Press. Part of the appendix is Extract 2 of the present book.

LORENTZ, H. A. (1904) *Proc. Acad. Sci. Amst.* **6**, 809 (Extract 3 of the present book).

LUDWIG, G. (1968) *Wave Mechanics*, Pergamon, Oxford.

MAXWELL, J. C. (1873) *Treatise on Electricity and Magnetism*, Oxford, Clarendon Press. (Maxwell made a rough sketch of the book some time during his time at King's College, London, 1860–5.)

MICHELSON, A. A. (1881) *Amer. Jour. Sci.* **22**, 20 (Extract 1 of the present book).

MICHELSON, A. A. and MORLEY, E. W. (1887) *Amer. Jour. Sci.* **34**, 333.

MILLER, D. C. (1925) *Proc. Nat. Acad. Sci.* **9**, 306.

MILNE, E. A. (1948) *Kinematic Relativity*, Oxford. (This book does use the method, but its systematic employment now owes much to H. Bondi.)

MINKOWSKI, H. (1908) *Göttinger Nachrichten*, p. 53. (But a clearer view is in his address to the 80th assembly of German physicists in Cologne, 1908, available in translation in *The Principle of Relativity*, A. EINSTEIN and others, Dover, New York (1923).)

NEWTON, I. S. (1786) *Philosophiae Naturalis Principia Mathematica*, London. (A. Motte's translation revised by F. Cajori (Cambridge, 1934) may be consulted.)

NOETHER, E. (1918) *Göttingen Nachrichten*, p. 235.

PLANCK, M. (1900) *Verh. D. Phys. Ges.* **2**, 202. (The reader who wishes to pursue the historical aspect of the discovery of quantum theory further should consult *The Old Quantum Theory*, D. TER HAAR, Pergamon Press, 1967.)

POINCARÉ, H. (1906) *Rend. del Circ. Mat. di Palermo* **21**, 129. (Extract 4 of the present book are pp. 129–46 and pp. 166–75.)

RAYLEIGH, Lord (1900) *Phil. Mag.* **49**, 539.

RUND, H. (1966) *The Hamilton–Jacobi Theory in the Calculus of Variations*, van Nostrand, London.

RUTHERFORD, E. (1911) *Phil. Mag.* **21**, 669.

SCHRÖDINGER, E. (1926) *Ann. der Phys.* **79**, 361.

SHANKLAND, R. S., McKUSKEY, S. W., LEONE, F. C. and KUERTI, G. (1955) *Rev. Mod. Phys.* **27**, 167.

THOMPSON, J. J. (1897) *Phil. Mag.* **44**, 293.

WEBER, W. and KOHLRAUSH, R. (1856) *Ann. der Phys.* **94**, 10.

WHITROW, G. J. (1961) *The Natural Philosophy of Time*, Nelson, London.

WIEN, W. (1893) *Ber. Berlin Akad. Wiss.* (9 Feb.).

WIEN, W. (1896) *Ann. der Phys.* **58**, 662.

WIGNER, E. (1939) *Ann. of Math.* **40**, 149 (Extract 10 of the present book).

WILSON, M. and WILSON, H. A. (1913) *Proc. Roy. Soc.* (A), **89**, 99 (Extract 6 of the present book).

ZEEMAN, P. (1914) *Proc. Acad. Sci. Amst.* **17**, 445 (Extract 7 of the present book).

PART II

NOTES ON EXTRACT 1

MICHELSON, in his fundamental paper about the velocity of the earth relative to the assumed medium of transmission of light, starts from Clerk Maxwell's letter to *Nature*, in which he points out that a measurement of the speed of light by means of the eclipses of the satellites of Jupiter would, in principle, be suitable for determining the velocity of the Earth through the medium of the velocity of light. In fact any such measurement would have to be much more accurate than those which can actually be made of eclipses, and Michelson therefore devises an apparatus, now known as the Michelson interferometer, for making this difference readily observable. A beam of light is split up at a half-silvered mirror and traverses two paths at right angles to each other, returning to the centre where the two beams of light are allowed to interfere. The arms are now rotated through 90°, and a watch is kept for any movement in the interferences fringes. If, for example, one arm were pointing in the direction of motion of the Earth through the aether at the beginning of the experiment, then when the second arm is transferred to that direction, the situation will be reversed, and any difference in times will show up as a shift of interference fringes.

EXTRACT 1[*]

Art. XXI. The Relative Motion of the Earth and the Luminiferous Ether

By ALBERT A. MICHELSON

Master, U.S. Navy

THE undulatory theory of light assumes the existence of a medium called the ether, whose vibrations produce the phenomena of heat and light, and which is supposed to fill all space. According to Fresnel, the ether, which is enclosed in optical media, partakes of the motion of these media, to an extent depending on their indices of refraction. For air, this motion would be but a small fraction of that of the air itself and will be neglected.

Assuming then that the ether is at rest, the earth moving through it, the time required for light to pass from one point to another on the earth's surface, would depend on the direction in which it travels.

Let V be the velocity of light.

v = the speed of the earth with respect to the ether.

D = the distance between the two points.

d = the distance through which the earth moves, while light travels from one point to the other.

d_1 = the distance earth moves, while light passes in the opposite direction.

Suppose the direction of the line joining the two points to coincide with the direction of earth's motion, and let T = time re-

[* *Amer. Jour. Sci.* **22,** 20 (1881).]

quired for light to pass from the one point to the other, and $T_1 =$ time required for it to pass in the opposite direction. Further, let $T_0 =$ time required to perform the journey if the earth were at rest.

Then $T = \dfrac{D+d}{V} = \dfrac{d}{v};$ and $T_1 = \dfrac{D-d}{V} = \dfrac{d_1}{v}.$

From these relations we find $d = Dv/(V-v)$ and $d_1 = Dv/(V+v)$ whence $T = D/(V-v)$ and $T_1 = D/(V+v)$; $T-T_1 = 2T_0(v/V)$ *nearly*, and $v = V(T-T_1)/2T_0$.

If now it were possible to measure $T-T_1$ since V and T_0 are known, we could find v the velocity of the earth's motion through the ether.

In a letter, published in "Nature" shortly after his death, Clerk Maxwell pointed out that $T-T_1$ could be calculated by measuring the velocity of light by means of the eclipses of Jupiter's satellites at periods when that planet lay in different directions from earth; but that for this purpose the observations of these eclipses must greatly exceed in accuracy those which have thus far been obtained. In the same letter it was also stated that the reason why such measurements could not be made at the earth's surface was that we have thus far no method for measuring the velocity of light which does not involve the necessity of returning the light over its path, whereby it would lose nearly as much as was gained in going.

The difference depending on the square of the ratio of the two velocities, according to Maxwell, is far too small to measure.

The following is intended to show that, with a wave-length of yellow light as a standard, the quantity—if it exists—is easily measurable.

Using the same notation as before we have $T = D/(V-v)$ and $T_1 = D/(V+v)$. The whole time occupied therefore in going and returning $T+T_1 = 2DV/(V^2-v^2)$. If, however, the light had traveled in a direction at right angles to the earth's motion it would be entirely unaffected and the time of going and returning would be, therefore, $2(D/V) = 2T_0$. The difference between the times $T+T_1$ and $2T_0$ is

$$2DV\left(\frac{1}{V^2-v^2} - \frac{1}{V^2}\right) = \tau; \qquad \tau = 2DV\frac{v^2}{V^2(V^2-v^2)}$$

or nearly $2T_0(v^2/V^2)$. In the time τ the light would travel a distance $V\tau = 2VT_0(v^2/V^2) = 2D(v^2/V^2)$.

That is, the actual distance the light travels in the first case is greater than in the second, by the quantity $2D(v^2/V^2)$.

Considering only the velocity of the earth in its orbit, the ratio $v/V = 1/10\,000$ approximately, and $v^2/V^2 = 1/100\,000\,000$. If $D = 1200$ millimeters, or in wave-lengths of yellow light, $2\,000\,000$, then in terms of the same unit, $2D(v^2/V^2) = 4/100$.

If, therefore, an apparatus is so constructed as to permit two pencils of light, which have traveled over paths at right angles to each other, to interfere, the pencil which has traveled in the direction of the earth's motion, will in reality travel $\frac{4}{100}$ of a wavelength farther than it would have done, were the earth at rest. The other pencil being at right angles to the motion would not be affected.

If, now, the apparatus be revolved through $90°$ so that the second pencil is brought into the direction of the earth's motion, its path will have lengthened $\frac{4}{100}$ wave-lengths. The total change in the position of the interference bands would be $\frac{8}{100}$ of the distance between the bands, a quantity easily measurable.

The conditions for producing interference of two pencils of light which had traversed paths at right angles to each other were realized in the following simple manner.

Light from a lamp a, fig. 1, passed through the plane parallel glass plate b, part going to the mirror c, and part being reflected to the mirror d. The mirrors c and d were of plane glass, and silvered on the front surface. From these the light was reflected to b, where the one was reflected and the other refracted, the two coinciding along be.

SPECIAL RELATIVITY

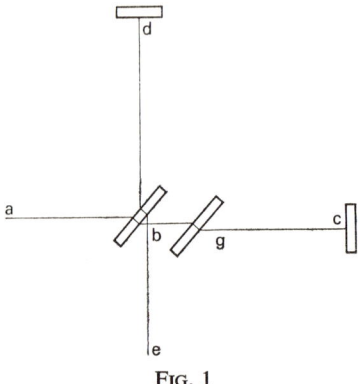

Fig. 1

The distance *bc* being made equal to *bd*, and a plate of glass *g* being interposed in the path of the ray *bc*, to compensate for the thickness of the glass *b*, which is traversed by the ray *bd*, the two rays will have traveled over equal paths and are in condition to interfere.

The instrument is represented in plan by fig. 2, and in perspective by fig. 3. The same letters refer to the same parts in the two figures.

The source of light, a small lantern provided with a lens, the flame being in the focus, is represented at *a*. *b* and *g* are the two plane glasses, both being cut from the same piece; *d* and *c* are the silvered glass mirrors; *m* is a micrometer screw which moves the plate *b* in the direction *bc*. The telescope *e*, for observing the interference bands, is provided with a micrometer eyepiece. *w* is a counterpoise.

In the experiments the arms, *bd*, *bc*, were covered by long paper boxes, not represented in the figures, to guard against changes in temperature. They were supported at the outer ends by the pins *k*, *l*, and at the other by the circular plate *o*. The adjustments were effected as follows:

The mirrors *c* and *d* were moved up as close as possible to the plate *b*, and by means of the screw *m* the distances between a point

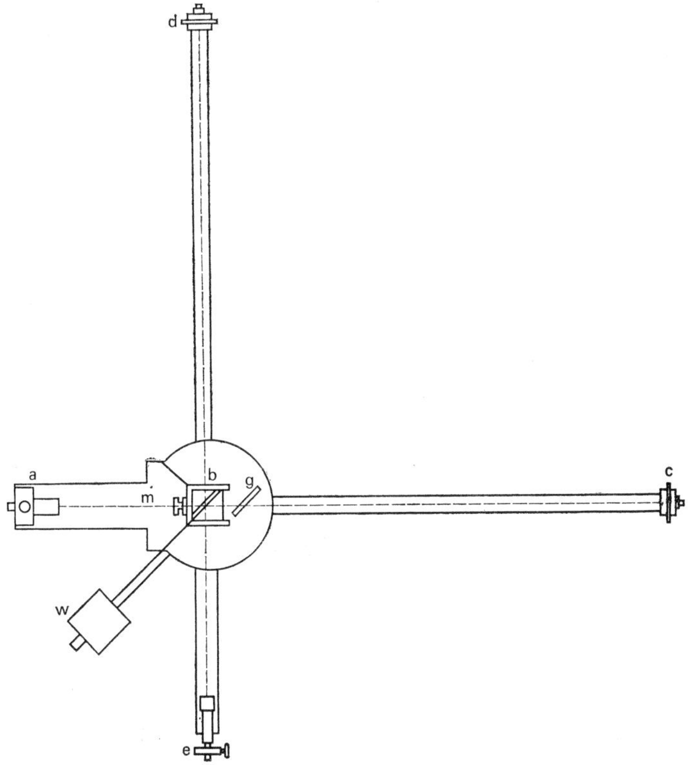

Fig. 2

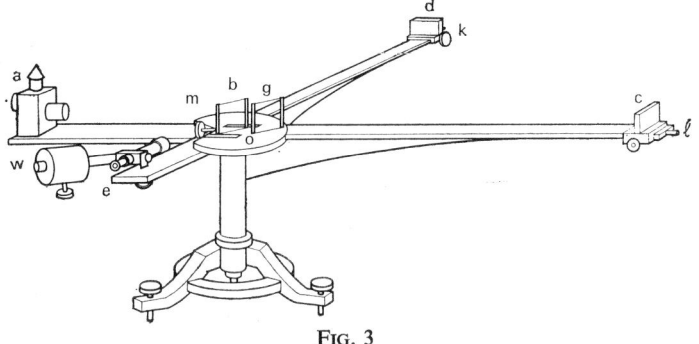

Fig. 3

on the surface of *b* and the two mirrors were made approximately equal by a pair of compasses. The lamp being lit, a small hole made in a screen placed before it served as a point of light; and the plate *b*, which was adjustable in two planes, was moved about till the two images of the point of light, which were reflected by the mirrors, coincided. Then a sodium flame placed at *a* produced at once the interference bands. These could then be altered in width, position, or direction, by a slight movement of the plate *b*, and when they were of convenient width and of maximum sharpness, the sodium flame was removed and the lamp again substituted. The screw *m* was then slowly turned till the bands reappeared. They were then of course colored, except the central band, which was nearly black. The observing telescope had to be focussed on the surface of the mirror *d*, where the fringes were most distinct. The whole apparatus, including the lamp and the telescope, was movable about a vertical axis.

It will be observed that this apparatus can very easily be made to serve as an "interferential refractor," and has the two important advantages of small cost, and wide separation of the two pencils.

The apparatus as above described was constructed by Schmidt and Haensch of Berlin. It was placed on a stone pier in the Physical Institute, Berlin. The first observation showed, however, that owing to the extreme sensitiveness of the instrument to vibrations, the work could not be carried on during the day. The experiment was next tried at night. When the mirrors were placed half-way on the arms the fringes were visible, but their position could not be measured till after twelve o'clock, and then only at intervals. When the mirrors were moved out to the ends of the arms, the fringes were only occasionally visible.

It thus appeared that the experiments could not be performed in Berlin, and the apparatus was accordingly removed to the *Astrophysicalisches Observatorium* in Potsdam. Even here the ordinary stone piers did not suffice, and the apparatus was again transferred, this time to a cellar whose circular walls formed the foundation for the pier of the equatorial.

Here, the fringes under ordinary circumstances were sufficiently quiet to measure, but so extraordinarily sensitive was the instrument that the stamping of the pavement, about 100 meters from the observatory, made the fringes disappear entirely!

If this was the case with the instrument constructed with a view to avoid sensitiveness, what may we not expect from one made as sensitive as possible!

At this time of the year, early in April, the earth's motion in its orbit coincides roughly in longitude with the estimated direction of the motion of the solar system—namely, toward the constellation Hercules. The direction of this motion is inclined at an angle of about $+26°$ to the plane of the equator, and at this time of the year the tangent of the earth's motion in its orbit makes an angle of $-23\frac{1}{2}°$ with the plane of the equator; hence we may say the resultant would lie within $25°$ of the equator.

The nearer the two components are in magnitude to each other, the more nearly would their resultant coincide with the plane of the equator.

In this case, if the apparatus be so placed that the arms point north and east at noon, the arm pointing east would coincide with the resultant motion, and the other would be at right angles. Therefore, if at this time the apparatus be rotated $90°$, the displacement of the fringes should be *twice* $\frac{8}{100}$ or $0·16$ of the distance between the fringes.

If, on the other hand, the proper motion of the sun is small compared to the earth's motion, the displacement should be $\frac{6}{10}$ of $·08$ or $0·048$. Taking the mean of these two numbers as the most probable, we may say that the displacement to be looked for is not far from one-tenth the distance between the fringes.

The principal difficulty which was to be feared in making these experiments, was that arising from changes of temperature of the two arms of the instrument. These being of brass whose coefficient of expansion is 0·000019 and having a length of about 1000 mm. or 1 700 000 wave-lengths, if one arm should have a temperature only one one-hundredth of a degree higher than the other, the fringes would thereby experience a displacement three times as great as that which would result from the rotation. On the other hand, since the changes of temperature are independent of the direction of the arms, if these changes were not too great their effect could be eliminated.

It was found, however, that the displacement on account of bending of the arms during rotation was so considerable that the instrument had to be returned to the maker, with instructions to make it revolve as easily as possible. It will be seen from the tables, that notwithstanding this precaution a large displacement was observed in one particular direction. That this was due entirely to the support was proved by turning the latter through 90°, when the direction in which the displacement appeared was also changed 90°.

On account of the sensitiveness of the instrument to vibration, the micrometer screw of the observing telescope could not be employed, and a scale ruled on glass was substituted. The distance between the fringes covered three scale divisions, and the position of the center of the dark fringe was estimated to fourths of a division, so that the separate estimates were correct to within $\frac{1}{12}$.

It frequently occurred that from some slight cause (among others the springing of the tin lantern by heating) the fringes would suddenly change their position, in which case the series of observations was rejected and a new series begun.

In making the adjustment before the third series of observations, the direction in which the fringes moved, on moving the glass plate b, was reversed, so that the displacement in the third and fourth series are to be taken with the opposite sign.

At the end of each series the support was turned 90°, and the axis was carefully adjusted to the vertical by means of the foot-screws and a spirit level.

The heading of the columns in the table gives the direction toward which the telescope pointed.

The footing of the erroneous column is marked x, and in the calculations the mean of the two adjacent footings is substituted.

The numbers in the columns are the positions of the center of the dark fringe in *twelfths* of the distance between the fringes.

In the first two series, when the footings of the columns N. and S. exceed those of columns E. and W., the excess is called positive. The excess of the footings of N.E., S.W., over those of N.W., S.E., are also called positive. In the third and fourth series this is reversed.

The numbers marked "excess" are the sums of ten observations. Dividing therefore by 10, to obtain the mean, and also by 12 (since the numbers are twelfths of the distance between the fringes), we find for

	N.S.	N.E., S.W.
Series 1	$+0.017$	$+0.050$
Series 2	-0.025	-0.033
Series 3	$+0.030$	$+0.030$
Series 4	$+0.067$	$+0.087$
	4) 0.089	0.137
Mean =	$+0.022$	$+0.034$

The displacement is, therefore,

In favor of the columns N.S. .... $+0.022$
In favor of the columns N.E., S.W. .... $+0.034$

The former is too small to be considered as showing a displacement due to the simple change in direction, and the latter should have been zero.

The numbers are simply outstanding errors of experiment. It is, in fact, to be seen from the footings of the columns, that the numbers increase (or decrease) with more or less regularity from left to right.

This gradual change, which should not in the least affect the periodic variation for which we are searching, would of itself ne-

100 SPECIAL RELATIVITY

	N.	N.E.	E.	S.E.	S.	S.W.	W.	NW.	Remarks.
1st revolution	0.0	0.0	0.0	−8.0	−1.0	−1.0	−2.0	−3.0	Series 1, footscrew marked B, toward East.
2d revolution	16.0	16.0	16.0	9.0	16.0	16.0	15.0	13.0	
3d revolution	17.0	17.0	17.0	10.0	17.0	16.0	16.0	17.0	
4th revolution	15.0	15.0	15.0	8.0	14.5	14.5	14.5	14.0	
5th revolution	13.5	13.5	13.5	5.0	12.0	13.0	13.0	13.0	
	61.5	61.5	61.5	x	58.5	58.5	56.5	54.0	
S.	58.5	W.	56.5		N.E.	61.5	S.E.	60.0	
	120.0		118.0			120.0		114.0	
	118.0					114.0			
Excess,	+2.0					+6.0			
1st revolution	10.0	11.0	12.0	13.0	13.0	0.0	14.0	15.0	Series 2, B toward South.
2d revolution	16.0	16.0	16.0	17.0	17.0	2.0	17.0	17.0	
3d revolution	17.5	17.5	17.5	17.5	17.5	4.0	18.0	17.5	
4th revolution	17.5	17.5	17.0	17.0	17.0	4.0	17.0	17.0	
5th revolution	17.0	17.0	17.0	17.0	16.0	3.0	16.0	16.0	
	78.0	79.0	79.5	81.5	80.5	x	82.0	82.5	
S.	80.5	W.	82.0		N.E.	79.0	S.E.	81.5	
	158.5		161.5			160.0		164.0	
	161.5					164.0			
Excess,	−3.0					−4.0			

MICHELSON: THE LUMINIFEROUS ETHER

Series 3, B toward West

1st revolution	3.0	3.0	3.0	3.0	2.5	2.5	2.5	10.0
2d revolution	18.0	17.5	17.5	18.0	18.5	19.0	19.5	26.0
3d revolution	11.0	11.0	13.0	12.0	13.0	13.5	13.5	21.0
4th revolution	1.0	0.0	0.5	0.5	0.5	0.0	0.0	14.0
5th revolution	4.0	4.0	5.0	5.0	5.0	5.5	5.5	16.0
	37.0	35.5	39.0	38.5	39.5	40.5	71.0	x
	39.5	W.	41.0		N.E.	35.5	S.E.	38.5
	76.5		80.0			76.0		79.5
	S.		76.5					76.0
Excess,			+3.5					+3.5

Series 4, B toward North.

1st revolution	14.0	21.0	15.5	17.0	14.0	14.5	14.5	16.0
2d revolution	10.0	20.0	12.0	12.0	13.0	13.0	13.0	13.5
3d revolution	14.0	25.0	15.0	16.0	16.0	16.0	16.0	17.0
4th revolution	18.0	27.0	18.5	18.5	18.5	19.0	20.0	21.0
5th revolution	15.0	24.0	15.0	15.0	15.0	16.0	16.0	16.5
	71.0	x	76.0	78.5	76.5	78.5	79.5	84.0
	76.5	W.	79.5		N.E.	73.5	S.E.	78.5
	147.5		155.5			152.0		162.5
	S.		147.5					152.0
Excess,			+8.0					+10.5

cessitate an outstanding error, simply because the sum of the two columns farther to the left must be less (or greater) than the sum of those farther to the right.

This view is amply confirmed by the fact that where the excess is positive for the column N.S., it is also positive for N.E., S.W., and where negative, negative. If, therefore, we can eliminate this gradual change, we may expect a much smaller error. This is most readily accomplished as follows:

Adding together all the footings of the four series, the third and fourth with negative sign, we obtain

N.	N.E.	E.	S.E.	S.	S.W.	W.	N.W.
31·5	31·5	26·0	24·5	23·0	20·8	18·0	11·0

or dividing by 20×12 to obtain the means in terms of the distance between the fringes,

N.	N.E.	E.	S.E.	S.	S.W.	W.	N.W.
0·131	0·131	0·108	0·102	0·096	0·086	0·075	0·046

If x is the number of the column counting from the right and y the corresponding footing, then the method of least squares gives as the equation of the straight line which passes nearest the points x, y—

$$y = 9 \cdot 25x + 64 \cdot 5$$

If, now, we construct a curve with ordinates equal to the difference of the values of y found from the equation, and the actual value of y, it will represent the displacements observed, freed from the error in question.

These ordinates are:

N.	N.E.	E.	S.E.	S.	S.W.	W.	N.W.
−·002	−·011	+·003	−·001	−·004	−·003	−·001	+·018

	N.		E.	S.E.	N.E.		N.W.	
	−·002		+·003		−·011		+·018	
S.	−·004	W.	−.001	S.W.	−·003	S.E.	−·001	
Mean =	−·003		+·001	Mean =	−·007		+·008	
	+·001				+·008			
Excess =	−·004			Excess =	−·015			

The small displacements $-0\cdot004$ and $-0\cdot015$ are simply errors of experiment.

The results obtained are, however, more strikingly shown by constructing the actual curve together with the curve that should have been found if the theory had been correct. This is shown in fig. 4.

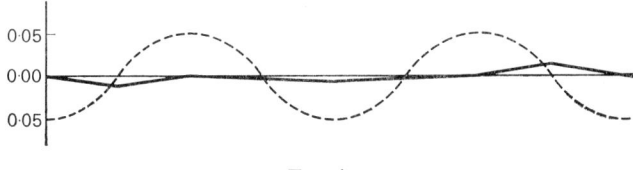

FIG. 4

The dotted curve is drawn on the supposition that the displacement to be expected is one-tenth of the distance between the fringes, but if this displacement were only $\frac{1}{100}$, the broken line would still coincide more nearly with the straight line than with the curve.

The interpretation of these results is that there is no displacement of the interference bands. The result of the hypothesis of a stationary ether is thus shown to be incorrect, and the necessary conclusion follows that the hypothesis is erroneous.

This conclusion directly contradicts the explanation of the phenomenon of aberration which has been hitherto generally accepted, and which presupposes that the earth moves through the ether, the latter remaining at rest.

It may not be out of place to add an extract from an article published in the Philosophical Magazine by Stokes in 1846.

"All these results would follow immediately from the theory of aberration which I proposed in the July number of this magazine; nor have I been able to obtain any result admitting of being compared with experiment, which would be different according to which theory we adopted. This affords a curious instance of two totally different theories running parallel to each other in the explanation of phenomena. I do not suppose that many would be disposed to maintain Fresnel's theory, when it is shown that it may be dispensed with, inasmuch as we would not be disposed to believe, without good evidence, that the ether moved quite freely through the solid mass of the earth. Still it would have been satisfactory, if it had been possible to have put the two theories to the test of some decisive experiment."

In conclusion, I take this opportunity to thank Mr. A. Graham Bell, who has provided the means for carrying out this work, and Professor Vogel, the Director of the *Astrophysicalisches Observatorium*, for his courtesy in placing the resources of his laboratory at my disposal.

NOTES ON EXTRACT 2

IN THE section of his book *Aether and Matter* reproduced here Larmor can be seen making great efforts to relate the field theory resulting from Maxwell's equations to the form of wave motion with which he is already familiar, that is, waves in elastic solids and fluids. From our point of view it no longer seems worth while constructing models of this kind, but the difficulty of doing so undoubtedly played a part in the disenchantment of physicists with the concept of the aether. With great ingenuity Larmor shows that the apparently paradoxical combinations of a perfect fluid (one having no viscosity) endowed with rotational elasticity can be realised by an immense aggregate of tiny gyroscopes. He then goes on to consider whether such a material could also provide a model for an electron in the sense of a point singularity. This part of the book is a quotation from an earlier paper in 1897, and is followed by a correction. This correction, to do with the constancy of the rotation, allows Larmor to carry the model a little further. It is, however, very striking that he comes to the conclusion that such a mechanical model is essentially only equivalent to constructing a variational principle for the field under consideration, a principle which, as he was well aware, can be constructed irrespective of the physical model. In the concluding pages of the extract he is anticipating to a considerable extent the later views of people about the utility of such models. (The extract consists of Appendix E, up to the middle of p. 334; the remainder of the appendix contains rather technical matter.)

EXTRACT 2[*]

APPENDIX E

On Kinematic and Mechanical Modes of Representation of the Activity of the Aether

By J. LARMOR

Mechanical Models and Illustrations

"Although the Gaussian aspect of the subject, which would simply assert that the primary atoms of matter exert actions on each other which are transmitted in time across space in accordance with Maxwell's equations, is a formally sufficient basis on which to construct physical theory, yet the question whether we can form a valid conception of a medium which is the seat of this transmission is of fundamental philosophical interest, quite independently of the fact that in default of the analogy at any rate of such a medium this theory would be too difficult for development. With a view to further assisting a judgment on this question, it is here proposed to describe a process by which a dynamical model of this medium can be theoretically built up out of ordinary matter,—not indeed a permanent model, but one which can be made to continue to represent the aether for any assignable finite time, though it must ultimately decay. The aether is a perfect fluid endowed with rotational elasticity; so in the first place we have—and this is the most difficult part of our undertaking—to construct a material model of a perfect fluid, which is a type of medium nowhere existing in the material world. Its

[* *Aether and Matter*, pp. 323–34.]

characteristics are continuity of motion and absence of viscosity: on the other hand in an ordinary fluid, continuity of motion is secured by diffusion of momentum by the moving molecules, which is itself viscosity, so that it is only in motions such as vibrations and slight undulations where the other finite effects of viscosity are negligible, that we can treat an ordinary fluid as a perfect one. If we imagine an aggregation of frictionless solid spheres, each studded over symmetrically with a small number of frictionless spikes (say four) of length considerably less than the radius*, so that there are a very large number of spheres in the differential element of volume, we shall have a possible though very crude means of representation of an ideal perfect fluid. There is next to

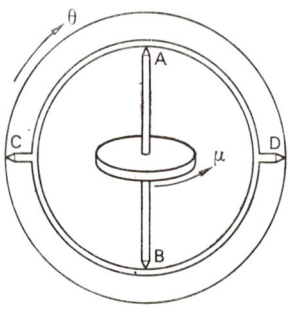

be imparted to each of these spheres the elastic property of resisting absolute rotation; and in this we follow the lines of Lord Kelvin's gyrostatic vibratory aether. Consider a gyrostat consisting of a flywheel spinning with angular momentum μ, with its axis AB pivoted as a diameter on a ring whose perpendicular diameter CD is itself pivoted on the sphere, which may for example be a hollow shell with the flywheel pivoted in its interior; and

---

* The use of these studs is to maintain continuity of motion of the medium without the aid of viscosity; and also (§ 4) to compel each sphere to participate in the rotation of the element of volume of the medium, so that the latter shall be controlled by the gyrostatic torques of the spheres.

examine the effect of imparting a small rotational displacement to the sphere. The direction of the axis of the gyrostat will be displaced only by that component of the rotation which is in the plane of the ring; an angular velocity $d\theta/dt$ in this plane will produce a torque measured by the rate of change of the angular momentum, and therefore by the parallelogram law equal to $\mu\, d\theta/dt$ turning the ring round the perpendicular axis CD, thus involving a rotation of the ring round that axis with angular acceleration $\mu/i.\, d\theta/dt$, that is with velocity $\mu/i.\,\theta$, where i is the aggregate moment of inertia of the ring and the flywheel about a diameter of the wheel. Thus when the sphere has turned through a small angle θ, the axis of the gyrostat will be turning out of the plane of θ with an angular velocity $\mu/i.\,\theta$, which will persist uniform so long as the displacement of the sphere is maintained. This angular velocity again involves, by the law of vector composition, a decrease of gyrostatic angular momentum round the axis of the ring at the rate $\mu^2/i.\,\theta$; accordingly the displacement θ imparted to the sphere originates a gyrostatic opposing torque, equal to $\mu^2/i.\,\theta$ so long as $\mu/i.\,\int \theta\, dt$ remains small, and therefore of purely elastic type. If then there are mounted on the sphere three such rings in mutually perpendicular planes, having equal free angular momenta associated with them, the sphere will resist absolute rotation in all directions with isotropic elasticity. But this result holds only so long as the total displacement of the axes of the flywheels is small: it suffices however to confer rotatory elasticity, as far as is required for the purpose of the transmission of vibrations of small displacement through a medium constituted of a flexible framework with such gyrostatic spheres attached to its links, which is Lord Kelvin's gyrostatic model* of the luminiferous working of the aether. For the present purpose we require this quality of perfect rotational elasticity to be permanently main-

* Lord Kelvin, *Comptes Rendus*, Sept. 1889: 'Math. and Phys. Papers,' III, p. 466.

108 SPECIAL RELATIVITY

tained, whether the disturbance is vibratory or continuous. Now observe that if the above associated free angular momentum μ is taken to be very great, it will require a proportionately long time for a given torque to produce an assigned small angular displacement, and this time we can thus suppose prolonged as much as we please: observe further that the motion of our rotational aether in the previous papers is irrotational except where electric force exists which produces rotation proportional to its intensity, and that we have been compelled to assume a high coefficient of inertia of the medium, and therefore an extremely high elasticity in order to conserve the ascertained velocity of radiation, so that the very strongest electric forces correspond to only very slight rotational displacements of the medium: and it follows that the arrangement here described, though it cannot serve as a model of a field of steady electric force lasting for ever, can yet theoretically represent such a field lasting without sensible decay for any length of time that may be assigned.

"It remains to attempt a model (cf. Part I, § 116) of the constitution of an electron, that is of one of the point-singularities in the uniform aether which are taken to be the basis of matter, and at any rate are the basis of its electrical phenomena. Consider the medium composed of studded gyrostatic spheres as above: although the motions of the aether, as distinct from the matter which flits across it, are so excessively slow on account of its great inertia that viscosity might possibly in any case be neglected, yet it will not do to omit the studs and thus make the model like a model of a gas, for we require rotation of an individual sphere to be associated with rotation of the whole element of volume of the medium in which it occurs. Let then in the rotationally elastic medium a narrow tubular channel be formed, say for simplicity a straight channel AB of uniform section: suppose the walls of this channel to be grasped, and rotated round the axis of the tube, the rotation at each point being proportional for the straight tube to $AP^{-2}+PB^{-2}$*: this rotation will be distributed through the medium, and as the result there will be lines of rotational displacement all starting from A and terminating at B: and so long as the walls of the channel are held in this position by extraneous force, A will be a positive electron in the medium, and B will be the complementary negative one. They will both disappear together when the walls of the channel are released. But now suppose that before this release the channel is filled up (except small vacuous nuclei at A and B which will assume the spherical form) with studded gyrostatic spheres so as to be continuous with the surrounding medium; the effort of release in this surrounding medium will rotate these spheres slightly until they attain the state of equilibrium in which the rotational elasticity of the new part of the medium formed by their aggregate provides a balancing torque, and the conditions all round A or B will finally be symmetrical. We shall thus have created two permanent conjugate electrons A and B; each of them can be moved about through the medium, but they will both persist until they are destroyed by an extraneous process the reverse of that by which they are formed. Such constraints as may be necessary to prevent division of their vacuous nuclei are outside our present scope; and mutual destruction of two complementary electrons by direct impact is an occurrence of infinitely small probability. The model of an electron thus formed will persist for any finite assignable time if the distribution of gyrostatic momentum in the medium is sufficiently intense: but the constitution of our model of the medium itself of course prevents, in this respect also, absolute permanence. It is not by any means here suggested that this circumstance forms any basis for speculation as to whether matter is permanent, or will gradually fade away. The position that we are concerned in supporting is that the cosmical theory which is used in the present memoirs as a descriptive basis for ultimate physical discus-

[* This is corrected *infra*.]

sions is a consistent and thinkable scheme; one of the most convincing ways of testing the possibility of the existence of any hypothetical type of mechanism being the scrutiny of a specification for the actual construction of a model of it.

"An idea of the nature and possibility of a self-locked intrinsic strain, such as that here described, may be facilitated by reference to the cognate example of a *material* wire welded into a ring after twist has been put into it. We can also have a closer parallel, as well as a contrast; if breach of continuity is produced across an element of interface in the midst of an incompressible medium endowed with *ordinary material rigidity*, for example by the creation of a lens-shaped cavity, and the material on one side of the breach is twisted round in its plane, and continuity is then restored by cementing the two sides together, a model of an electric doublet or polar molecule will be produced, the twist in the medium representing the electric displacement and being at a distance expressible as due to two conjugate poles in the ordinary manner. Such a doublet is permanent, as above; it can be displaced into a different position, at any distance, as a strain-form, without the medium moving along with it; such displacement is accompanied by an additional strain at each point in the medium, namely, that due to the doublet in its new position together with a negative doublet in the old one. A series of such doublets arranged transversely round a linear circuit will represent the integrated effect of an electric polarization-current in that circuit; they will imply irrotational linear displacement of the medium round the circuit after the manner of vortex motion, but this will now involve elastic stress on account of the rigidity. Thus with an ordinary elastic solid medium, the phenomena of dielectrics, including wave-propagation, may be kinematically illustrated; but we can thereby obtain no representation of a single isolated electric charge or of a current of conduction, and the laws of optical reflexion would be different from the actual ones. This material lilustration will clearly extend to the dynamical laws of induction and electromagnetic attraction between alternating currents, but only in so far as they are derived from the kinetic energy; the law of static attraction between doublets of this kind would be different from the actual electric law."

Phil. Trans. 1897 A, pp. 209–212.

1. This description of an ideal (supernatural) construction for electrons in a rotational aether requires correction as regards one point. The line integral of the rotation that has to be imparted to the walls of the canal AB is equal at each cross-section to the surface integral of the normal component of the rotational displacement of the aether over a surface abutting on it and enclosing either electron: it is therefore constant all along the canal, whether the latter is straight or curved, instead of proportional to $AP^{-2}+PB^{-2}$ as above stated. Thus if the canal is of uniform circular section, the rotational displacement of its walls is uniform all along it.

This circumstance allows a development of the analogy, which will further illustrate the origin of the mechanical attraction between two electrons. It is a well-known device in mechanical construction, to use a flexible wire of great torsional rigidity to transmit rotation from one shaft to another not in line with it, by clamping the ends of the wire to the ends of the shafts so that it forms an elastic connexion between them. Now instead of filling up our ideal canal in the aether by a filament of aether, let us suppose it filled up by such a wire, of infinite torsional rigidity, and in continuous connexion with the surrounding aether. Each time any cross section C of this wire is rotated round its axis by an impressed torque, the rotation is transmitted all along the wire, and thence to the aether alongside it; and two complementary electrons are thus developed at its ends A and B. On releasing this section C the rotation undoes itself, and these terminal electrons disappear. This arrangement constitutes an elastic system

112 SPECIAL RELATIVITY

devoid of any intrinsic stress such as was previously implanted in the system by filling up the canal with aether itself; for it becomes free from stress on releasing the wire. We should therefore be in a position to point directly to the proximate cause of the attraction of one electron on the other. It is to be found in the tangential tractions which the surrounding aether exerts on the surface of the wire, which form a system of forces statically equivalent, by virtue of the principle of virtual work, to an attraction between its ends.

We can in this way imagine the aether with its contained electrons as mathematically dissected into an elastic medium devoid of intrinsic strain, by connecting each positive electron with a complementary negative one by means of such an elastic material wire AB in continuous connexion with the aether, to which has been imparted at any cross section C the amount of rotation proper to maintain the intensities of the electrons. When the wire has disappeared and the electrons at A and B are permanently constituted by filling up its place with aether, the possibility of thus specifying a proximate cause of the mechanical attraction between the electrons has also in a sense disappeared. But just as the exploration of the relations of a cyclic analytical function requires the introduction of cross-cuts or barriers in its domain to render account of the cyclic character, so the complete elucidation of the dynamics of a medium involving cyclic intrinsic strain requires the introduction of ideal canals or tubes connecting the strain-centres, through operation on which this strain may be considered as implanted in the medium*. We can even consider

* When the medium is thus completely specified, the line integration in Stokes' theorem of curl will contain integrals round the sections of these tubes where they cross the sheet. But it is only the ends of the tubes that are determinate; hence to obtain a definite result we must (as in i, p. 90, in which the fluxion dots should be deleted) apply the theorem only to the change, indicated there by the Δ, that results from small displacements of the existing electrons; each displacement of an electron is formally equivalent to the establishment of a tube of strain connecting its old with its new position, and

the tractions exerted on the surface of such a tube of strain as statically transmitted to the electrons at its ends just as if it included the wire of the illustration. Even when the wire is present the amount of the attraction is most easily determined by application of the principle of Energy: this method remains available when it is absent, so long as it is definitely recognized that the Energy principle, or more generally the Action principle, is a fundamental dynamical method whose application is not limited to the class of cases in which we are able to *describe* the activity of the medium in terms of familiar processes of direct elastic transmission. Although the simultaneous representation of the two kinds of existing forcive, aethereal stress and material attractions, thus transcends the usual elementary notion of elastic propagation, they yet appear alongside each other in the development of the dynamical formulation of the medium in terms of the principle of Action, which is prior to any model whatever, and is moreover logically required, unless we are content to view the medium as a system of relations in space and time represented by differential equations devoid of dynamical significance. We thus conclude, along with von Helmholtz, that there is no resting place in general dynamical theory or explanation, short of the Action foundation. The content of this principle, as applied to continuous media, is in various ways wider than the conception of simple elastic transmission, which is the case that is most familiar in the more easily analyzed classes of physical phenomena. We might for example have an energy function involving

---

whenever this tube crosses the sheet a correcting term is required in the formula.

A convenient mode of developing the electrodynamics of material media would be to replace the translational displacement of each electron by the local rotational displacement of the aether itself which is its constitutive equivalent as regards that medium; the problem can then be treated by methods of continuous analysis applied to free aether. Cf. *Camb. Phil. Trans.*, Stokes' Jubilee volume, 1900.

second as well as first differential coefficients of the displacements, in which case disturbances would still be transmitted by the medium, but not by the agency of simple elastic stress definable in terms of surface tractions alone: it is only the extreme shortness of the range of molecular action compared with the size of the element of mass that is just sensible to our powers of observation, that debars this case from being a practical one.

In point of history, the dynamics of elastic propagation was first developed in a somewhat inexact way by Navier and Poisson, and attempts were subsequently made to establish it on an incomplete molecular foundation by Cauchy and others. But there was no reliable foothold obtainable even for this simple case until Green, by one of his strokes of genius, summarily included the whole matter under the Action principle. Reference to a transmitting medium was previously instructive by way of general illustration, for example in physical optics, but before the use of this principle by Green and by MacCullagh there was no sufficiently exact and general formulation of its possible modes of activity. It is in this way that the Action principle is prior even to the exact development of a theory of simple elastic transmission: and it is thus not surprising that it forms the most suitable basis when the transmitting medium is constituted in a more complex manner.

2. The subjects discussed in this book have in the main been treated without any hypothesis as to the structure of the nucleus of an electron. In a preliminary stage of the development of this theory, the analogy of an electron to a conductor carrying an electric charge suggested that the nucleus of an electron might be treated as a minute spherical region in which the aether is effectively devoid of elasticity: but this is not an essential or even probable feature. The illustration above given, of a nucleus of intrinsic strain in an elastic solid, indicates that what is essential is the concentration of 'beknottedness' in the small volume of the me-

dium which constitutes the nucleus, which would thus correspond to a small volume-electrification. Such an intrinsic strain-form is mobile through the medium, without thereby originating any new distribution of stress around it, because it is only rotation and not deformation of the aether that calls out elastic reaction; and this free mobility is an essential element in the theory. But the analysis into independent strains and rotations, on which it rests, requires that both strain and rotation shall be very small; thus the inertia of the medium must be very great, and each nucleus must be so constituted that the intrinsic rotations involved in its structure are so small that they can everywhere be treated as differential rotations, which is demanded by the linearity of the scheme of equations as well as by the mobility of the nucleus.

The dynamical scheme developed in Chapter VI is however based solely on the application of the method of Action to a medium uniform throughout all space, specified by the Lagrangian function $T-W$, and involving in its constitution mobile poles or electrons which by their aggregation form a representation of matter, at any rate in those respects in which it interacts with the aether. In that scheme the effective aethereal displacement represented by (ξ, η, ζ) need not be defined: it is not necessary (and it was not there intended) to assume it to be a translational displacement. The scheme thus stands on a formally definite basis independently of any knowledge of the type of disturbance that (ξ, η, ζ) represents: and it has not as yet been shown to be too narrow to represent the field of general physical actions.

In the model or illustration of the working of the aether which has been here described, this disturbance (ξ, η, ζ) is taken to represent translational displacement of the element of aether originally situated at the point (x, y, z). The medium is then one whose elasticity is purely and solely rotational. One object of the gyrostatic mode of representation above explained is to render the idea of rotational elasticity more familiar and more easily grasped, by illustrating it from the properties of an actual medium which could theoretically be constructed from ordinary matter. It is also of use towards allaying scruples that naturally arise as to the legitimacy of assuming a set of abstract properties of a type not met with in matter under ordinary conditions, and therefore liable to the suspicion of being somehow self-contradictory or in opposition to formally necessary dynamical principles: but though an actual model of such a medium forms a valuable and forcible illustration, the argument is logically complete without it. Such a gyrostatic model has no claim to be more than an illustration of the properties of the aether, for an aether of the present type can hardly on any scheme be other than a medium, or mental construction if that term is preferred, prior to matter and therefore not expressible in terms of matter.

This more special hypothesis that takes the variable (ξ, η, ζ) in p. 84 to be proportional to actual translational displacement, involves on the other hand a question of direct fact, as to which there are physical means of inquiry: its further consideration is therefore called for. It has been explained that, whatever be the character of the vector (ξ, η, ζ), the facts as regards the influence of the Earth's motion on optical phenomena, as well as the linear character (p. 96) of the electrodynamic equations, require that the aether shall be practically stagnant. On the present hypothesis this vector, whose time-gradient represents magnetic force, must therefore be equal to the translational displacement of the medium multiplied by a very large numerical constant. There is in fact no phenomenon known which is inconsistent with the ultimate simplification of passing analytically towards a limit, by taking the translational displacement to be indefinitely small and this multiplier indefinitely great.

The question suggests itself, as to what inducement there is to specify (f, g, h) as of the type of rotational displacement at all, seeing that the theory develops itself without any reference to the type of disturbance which this vector represents. The only motive is that the number of unconnected hypotheses, which dynamically cannot be independent, is thereby reduced: the possibility of the intrinsic elastic structure of an electron, and that of its free mobility, will be in the more indeterminate theory two new assumptions, both of unaccustomed character: while on the more special view they are both merged as corollaries in the single interpretation of the relations of the aethereal medium, so that the scheme proceeds on that basis alone. But in the case of a mind to which this simplification does not appeal, either as an elimination of a group of hypotheses that cannot from the nature of the case be independent and are so liable to the possibility of being inconsistent with each other, or else as an assistance to vivid apprehension of the relations*, the argument can proceed without any necessity for its adoption.

---

* It is desirable to further emphasize that these representations are illustrative, not essential: it may be held that they are too imperfect to be useful, without giving up anything essential in the theoretical formulation of the phenomena. In ultimate logic any physical representation is in fact a mental construction or analogy, designed to relieve the mind from the intangible and elusive character of a complex of abstract relations. It thus involves a correlation of a range of phenomena with something else that can be constructed either actually or mentally. It is however unreasonable to suppose that two things not the same can have *complete* identity of relations: on the other hand the universal employment of such ideal pictures constitutes evidence that they are legitimate and powerful aids to knowledge. Our mental image, whether abstract or illuminated by a model, cannot ever be completely identical with the complex of phenomena which it represents, though it is capable of continued approximation thereto. The essential problem is to determine in each case how deep the correspondence extends: if it is found to extend into unforeseen properties and lead to the recognition or prediction of new relations in the field of the actual phenomena, its propriety within due restrictions is usually considered to be vindicated: it is in fact in this way that most advances of knowledge arise. Cf. Hertz's *Mechanik*, Introduction.

NOTES ON EXTRACT 3

LORENTZ begins by recalling Michelson's well-known experiment, which he and Fitzgerald had explained provisionally by supposing that solid objects were slightly altered by their motion through the aether. This explanation is very difficult to maintain in the face of later experiments. Lorentz is also strongly impressed by Poincaré's objection that the explanation of Michelson's result by means of a new hypothesis is worthless if it is allowed to serve as a precedent, with a new hypothesis every time new experimental facts arise. Accordingly Lorentz carries out a new investigation ostensibly on the basis of his theory of electrons. In fact, however, the character of the electrons is unimportant. The investigation is essentially of how Maxwell's equations can be transformed by a transformation which agrees as far as possible with the one used in Newtonian mechanics, and yet retain their form. How far Lorentz was from a deeper understanding of the formula which he found can be seen by the equations (4) and (5) of his paper, where the relationship between the time and the old coordinates is slightly different from the one now considered. It was, however, an astonishing achievement to be able to carry out the whole investigation on the basis of the invariance of Maxwell's equations. (The tables referred to at the end of the extract (which is nearly but not completely the end of Lorentz' paper) have been omitted as now of little interest.)

EXTRACT 3[*]

Electromagnetic Phenomena in a System Moving with any Velocity less than that of Light

By H. A. LORENTZ

§ 1. THE problem of determining the influence exerted on electric and optical phenomena by a translation, such as all systems have in virtue of the Earth's annual motion, admits of a comparatively simple solution, so long as only those terms need be taken into account, which are proportional to the first power of the ratio between the velocity of translation v and the velocity of light c. Cases in which quantities of the second order, i.e. of the order v^2/c^2, may be perceptible, present more difficulties. The first example of this kind is Michelson's well-known interference-experiment, the negative result of which has led Fitzgerald and myself to the conclusion that the dimensions of solid bodies are slightly altered by their motion through the ether.

Some new experiments, in which a second order effect was sought for, have recently been published. Rayleigh* and Brace[†] have examined the question whether the Earth's motion may cause a body to become doubly refracting. At first sight this might be expected, if the just mentioned change of dimensions is admitted. Both physicists, however, have obtained a negative result.

In the second place Trouton and Noble[‡] have endeavoured to

[* *Proc. Acad. Sci. Amst.* **6,** 809 (1904).]
* Rayleigh, *Phil. Mag.* (6), **4,** 1902, p. 678.
† Brace, *Phil. Mag.* (6), **7,** 1904, p. 317.
‡ Trouton and Noble, *Phil. Trans. Roy. Soc. Lond.*, A **202.** 1903, p. 165.

detect a turning couple acting on a charged condenser, the plates of which make a certain angle with the direction of translation. The theory of electrons, unless it be modified by some new hypothesis, would undoubtedly require the existence of such a couple. In order to see this, it will suffice to consider a condenser with ether as dielectric. It may be shown that in every electrostatic system, moving with a velocity **v**,* there is a certain amount of "electromagnetic momentum." If we represent this, in direction and magnitude, by a vector **G**, the couple in question will be determined by the vector product[†]

$$[\mathbf{G}.\mathbf{v}]. \tag{1}$$

Now, if the axis of z is chosen perpendicular to the condenser plates, the velocity **v** having any direction we like; and if U is the energy of the condenser, calculated in the ordinary way, the components of **G** are given[‡] by the following formulae, which are exact up to the first order,

$$G_x = \frac{2U}{c^2} v_x, \quad G_y = \frac{2U}{c^2} v_y, \quad G_z = 0.$$

Substituting these values in (1), we get for the components of the couple, up to terms of the second order,

$$\frac{2U}{c^2} v_y v_z, \quad -\frac{2U}{c^2} v_x v_z, \quad 0.$$

These expressions show that the axis of the couple lies in the plane of the plates, perpendicular to the translation. If a is the angle between the velocity and the normal to the plates, the mo-

---

* A vector will be denoted by a Clarendon letter, its magnitude by the corresponding Latin letter.

† See my article: "Weiterbildung der Maxwell'schen Theorie. Electronentheorie," *Mathem. Encyclopädie*, V, 14, § 21, a. (This article will be quoted as *M.E.*)

‡ *M.E.*, § 56, c.

ment of the couple will be $U(v/c)^2 \sin 2a$; it tends to turn the condenser into such a position that the plates are parallel to the Earth's motion.

In the apparatus of Trouton and Noble the condenser was fixed to the beam of a torsion-balance, sufficiently delicate to be deflected by a couple of the above order of magnitude. No effect could however be observed.

§ 2. The experiments of which I have spoken are not the only reason for which a new examination of the problems connected with the motion of the Earth is desirable. Poincaré* has objected to the existing theory of electric and optical phenomena in moving bodies that, in order to explain Michelson's negative result, the introduction of a new hypothesis has been required, and that the same necessity may occur each time new facts will be brought to light. Surely this course of inventing special hypotheses for each new experimental result is somewhat artificial. It would be more satisfactory if it were possible to show by means of certain fundamental assumptions and without neglecting terms of one order of magnitude or another, that many electromagnetic actions are entirely independent of the motion of the system. Some years ago, I already sought to frame a theory of this kind.† I believe it is now possible to treat the subject with a better result. The only restriction as regards the velocity will be that it be less than that of light.

§ 3. I shall start from the fundamental equations of the theory of electrons.‡ Let **D** be the dielectric displacement in the ether, **H** the magnetic force, ϱ the volume-density of the charge of an electron, **v** the velocity of a point of such a particle, and **F** the ponderomotive force, i.e. the force, reckoned per unit charge, which is exerted by the ether on a volume-element of an electron. Then, if we use a

---

* Poincaré, *Rapports du Congrès de physique de 1900*, Paris, 1, pp. 22, 23.
† Lorentz, *Zittingsverslag Akad. v. Wet.*, **7**, 1899, p. 507; Amsterdam Proc., 1898–99, p. 427.
‡ *M.E.* § 2.

fixed system of co-ordinates,

$$\left.\begin{array}{c} \text{div } \mathbf{D} = \varrho, \quad \text{div } \mathbf{H} = 0, \\ \text{curl } \mathbf{H} = \dfrac{1}{c}\left(\dfrac{\partial \mathbf{D}}{\partial t} + \varrho \mathbf{v}\right), \\ \text{curl } \mathbf{D} = -\dfrac{1}{c}\dfrac{\partial \mathbf{H}}{\partial t}, \\ \mathbf{F} = \mathbf{D} + \dfrac{1}{c}\,[\mathbf{v}.\mathbf{H}]. \end{array}\right\} \quad (2)$$

I shall now suppose that the system as a whole moves in the direction of x with a constant velocity v, and I shall denote by $\mathbf{u}$ any velocity which a point of an electron may have in addition to this, so that

$$v_x = v + u_x, \quad v_y = u_y, \quad v_z = u_z.$$

If the equations (2) are at the same time referred to axes moving with the system, they become

$$\text{div } \mathbf{D} = \varrho, \quad \text{div } \mathbf{H} = 0,$$

$$\frac{\partial H_z}{\partial y} - \frac{\partial H_y}{\partial z} = \frac{1}{c}\left(\frac{\partial}{\partial t} - v\frac{\partial}{\partial x}\right)D_x + \frac{1}{c}\varrho(v+u_x),$$

$$\frac{\partial H_x}{\partial z} - \frac{\partial H_z}{\partial x} = \frac{1}{c}\left(\frac{\partial}{\partial t} - v\frac{\partial}{\partial x}\right)D_y + \frac{1}{c}\varrho u_y,$$

$$\frac{\partial H_y}{\partial x} - \frac{\partial H_x}{\partial y} = \frac{1}{c}\left(\frac{\partial}{\partial t} - v\frac{\partial}{\partial x}\right)D_z + \frac{1}{c}\varrho u_z,$$

$$\frac{\partial D_z}{\partial y} - \frac{\partial D_y}{\partial z} = -\frac{1}{c}\left(\frac{\partial}{\partial t} - v\frac{\partial}{\partial x}\right)H_x,$$

$$\frac{\partial D_x}{\partial z} - \frac{\partial D_z}{\partial x} = -\frac{1}{c}\left(\frac{\partial}{\partial t} - v\frac{\partial}{\partial x}\right)H_y,$$

$$\frac{\partial D_y}{\partial x} - \frac{\partial D_x}{\partial y} = -\frac{1}{c}\left(\frac{\partial}{\partial t} - v\frac{\partial}{\partial x}\right)H_z,$$

$$F_x = D_x + \frac{1}{c}(u_y H_z - u_z H_y),$$

$$F_y = D_y - \frac{1}{c} v H_z + \frac{1}{c}(u_z H_x - u_x H_z),$$

$$F_z = D_z + \frac{1}{c} v H_y + \frac{1}{c}(u_x H_y - u_y H_x).$$

§ 4. We shall further transform these formulae by a change of variables. Putting

$$\frac{c^2}{c^2 - v^2} = \beta^2, \tag{3}$$

and understanding by l another numerical quantity, to be determined further on, I take as new independent variables

$$x' = \beta l x, \quad y' = l y, \quad z' = l z, \tag{4}$$

$$t' = \frac{l}{\beta} t - \beta l \frac{v}{c^2} x, \tag{5}$$

and I define two new vectors $\mathbf{D}'$ and $\mathbf{H}'$ by the formulae

$$D'_x = \frac{1}{l^2} D_x, \quad D'_y = \frac{\beta}{l^2}\left(D_y - \frac{v}{c} H_z\right), \quad D'_z = \frac{\beta}{l^2}\left(D_z + \frac{v}{c} H_y\right),$$

$$H'_x = \frac{1}{l^2} H_x, \quad H'_y = \frac{\beta}{l^2}\left(H_y + \frac{v}{c} D_z\right), \quad H'_z = \frac{\beta}{l^2}\left(H_z - \frac{v}{c} D_y\right),$$

for which, on account of (3), we may also write

$$\left.\begin{array}{l} D_x = l^2 D'_x, \quad D_y = \beta l^2\left(D'_y + \frac{v}{c} H'_z\right), \quad D_z = \beta l^2\left(D'_z - \frac{v}{c} H'_y\right) \\ H_x = l^2 H'_x, \quad H_y = \beta l^2\left(H'_y - \frac{v}{c} D'_z\right), \quad H_z = \beta l^2\left(H'_z + \frac{v}{c} D'_y\right) \end{array}\right\}. \tag{6}$$

As to the coefficient l, it is to be considered as a function of v, whose value is 1 for $v = 0$, and which, for small values of v, differs from unity no more than by a quantity of the second order.

124   SPECIAL RELATIVITY

The variable t' may be called the "local time"; indeed, for $\beta = 1, l = 1$ it becomes identical with what I formerly denoted by this name.

If, finally, we put

$$\frac{1}{\beta l^3} \varrho = \varrho' \tag{7}$$

$$\beta^2 u_x = u'_x, \quad \beta u_y = u'_y, \quad \beta u_z = u'_z, \tag{8}$$

these latter quantities being considered as the components of a new vector $\mathbf{u}'$, the equations take the following form:

$$\left.\begin{aligned}
\operatorname{div}' \mathbf{D}' &= \left(1 - \frac{v u'_x}{c^2}\right) \varrho', \quad \operatorname{div}' \mathbf{H}' = 0, \\
\operatorname{curl}' \mathbf{H}' &= \frac{1}{c}\left(\frac{\partial \mathbf{D}'}{\partial t'} + \varrho' \mathbf{u}'\right), \\
\operatorname{curl}' \mathbf{D}' &= -\frac{1}{c} \frac{\partial \mathbf{H}'}{\partial t'},
\end{aligned}\right\} \tag{9}$$

$$\left.\begin{aligned}
F_x &= l^2 \left\{ D'_x + \frac{1}{c}(u'_y H'_z - u'_z H'_y) + \frac{v}{c^2}(u'_y D'_y + u'_z D'_z) \right\}, \\
F_y &= \frac{l^2}{\beta} \left\{ D'_y + \frac{1}{c}(u'_z H'_x - u'_x H'_z) - \frac{v}{c^2} u'_x D'_y \right\}, \\
F_z &= \frac{l^2}{\beta} \left\{ D'_z + \frac{1}{c}(u'_x H'_y - u'_y H'_x) - \frac{v}{c^2} u'_x D'_z \right\}.
\end{aligned}\right\} \tag{10}$$

The meaning of the symbols div' and curl' in (9) is similar to that of div and curl in (2); only, the differentiations with respect to x, y, z are to be replaced by the corresponding ones with respect to x', y', z'.

§ 5. The equations (9) lead to the conclusion that the vectors $\mathbf{D}'$ and $\mathbf{H}'$ may be represented by means of a scalar potential ϕ' and a vector potential **A**′. These potentials satisfy the equations*

$$\nabla'^2 \phi' - \frac{1}{c^2} \frac{\partial^2 \phi'}{\partial t'^2} = -\varrho' \tag{11}$$

$$\nabla'^2 A' - \frac{1}{c^2} \frac{\partial^2 \mathbf{A}'}{\partial t'^2} = -\frac{1}{c} \varrho' \mathbf{u}', \tag{12}$$

and in terms of them **D**′ and **H**′ are given by

$$\mathbf{D}' = -\frac{1}{c} \frac{\partial \mathbf{A}'}{\partial t'} - \text{grad}' \, \phi' + \frac{v}{c} \, \text{grad}' \, A'_x \tag{13}$$

$$\mathbf{H}' = \text{curl}' \, \mathbf{A}'. \tag{14}$$

The symbol ∇'^2 is an abbreviation for $\partial^2/\partial x'^2 + \partial^2/\partial y'^2 + \partial^2/\partial z'^2$, and grad′ ϕ' denotes a vector whose components are

$$\frac{\partial \phi'}{\partial x'}, \quad \frac{\partial \phi'}{\partial y'}, \quad \frac{\partial \phi'}{\partial z'}.$$

The expression grad′ A'_x has a similar meaning.

In order to obtain the solution of (11) and (12) in a simple form, we may take x', y', z' as the co-ordinates of a point P' in a space S', and ascribe to this point, for each value of t', the values of ϱ', **u**′, ϕ', **A**′, belonging to the corresponding point $P(x, y, z)$ of the electromagnetic system. For a definite value t' of the fourth independent variable, the potentials ϕ' and **A**′ at the point P of the system or at the corresponding point P' of the space S', are given by[†]

$$\phi' = \frac{1}{4\pi} \int \frac{[\varrho']}{r'} \, dS' \tag{15}$$

$$\mathbf{A}' = \frac{1}{4\pi c} \int \frac{[\varrho' \mathbf{u}']}{r'} \, dS'. \tag{16}$$

Here dS' is an element of the space S', r' its distance from P', and the brackets serve to denote the quantity ϱ' and the vector $\varrho' \mathbf{u}'$

---
* *M.E.*, §§ 4 and 10.
† *Ibid.*, §§ 5 and 10.

such as they are in the element dS', for the value $t'-r'/c$ of the fourth independent variable.

Instead of (15) and (16) we may also write, taking into account (4) and (7),

$$\phi' = \frac{1}{4\pi} \int \frac{[\varrho]}{r} dS \tag{17}$$

$$\mathbf{A}' = \frac{1}{4\pi c} \int \frac{[\varrho \mathbf{u}]}{r} dS, \tag{18}$$

the integrations now extending over the electromagnetic system itself. It should be kept in mind that in these formulae r' does not denote the distance between the element dS and the point (x, y, z) for which the calculation is to be performed. If the element lies at the point (x_1, y_1, z_1), we must take

$$r' = l\sqrt{[\beta^2(x-x_1)^2 + (y-y_1)^2 + (z-z_1)^2]}.$$

It is also to be remembered that, if we wish to determine ϕ' and $\mathbf{A}'$ for the instant at which the local time in P is t', we must take ϱ and $\varrho\mathbf{u}'$, such as they are in the element dS at the instant at which the local time of that element is $t'-r'/c$.

§ 6. It will suffice for our purpose to consider two special cases. The first is that of an electrostatic system, i.e. a system having no other motion but the translation with the velocity v. In this case $\mathbf{u}' = 0$, and therefore, by (12), $\mathbf{A}' = 0$. Also, ϕ' is independent of t', so that the equations (11), (13), and (14) reduce to

$$\left. \begin{array}{l} \nabla'^2 \phi' = -\varrho', \\ \mathbf{D}' = -\operatorname{grad}' \phi', \\ \mathbf{H}' = 0. \end{array} \right\} \tag{19}$$

After having determined the vector $\mathbf{D}'$ by means of these equations, we know also the ponderomotive force acting on electrons that belong to the system. For these the formulae (10) become, since $\mathbf{u}' = 0$,

$$F_x = l^2 D'_x, \quad F = \frac{l^2}{\beta} D'_y, \quad F_z = \frac{l^2}{\beta} D'_z. \tag{20}$$

The result may be put in a simple form if we compare the moving system Σ, with which we are concerned, to another electrostatic system Σ' which remains at rest, and into which Σ is changed if the dimensions parallel to the axis of x are multiplied by βl, and the dimensions which have the direction of y or that of z, by l—a deformation for which $(\beta l, l, l)$ is an appropriate symbol. In this new system, which we may suppose to be placed in the above-mentioned space S', we shall give to the density the value ϱ', determined by (7), so that the charges of corresponding elements of volume and of corresponding electrons are the same in Σ and Σ'. Then we shall obtain the forces acting on the electrons of the moving system Σ, if we first determine the corresponding forces in Σ', and next multiply their components in the direction of the axis of x by l^2, and their components perpendicular to that axis by l^2/β. This is conveniently expressed by the formula

$$\mathbf{F}(\Sigma) = \left(l^2, \frac{l^2}{\beta}, \frac{l^2}{\beta} \right) \mathbf{F}(\Sigma'). \tag{21}$$

It is further to be remarked that, after having found $\mathbf{D}'$ by (19), we can easily calculate the electromagnetic momentum in the moving system, or rather its component in the direction of the motion. Indeed, the formula

$$\mathbf{G} = \frac{1}{c} \int [\mathbf{D} \cdot \mathbf{H}] \, dS$$

shows that

$$G_x = \frac{1}{c} \int (D_y H_z - D_z H_y) \, dS.$$

Therefore, by (6), since $\mathbf{H}' = 0$

$$G_x = \frac{\beta^2 l^4 v}{c^2} \int (D_y'^2 + D_z'^2) \, dS = \frac{\beta l v}{c^2} \int (D_y'^2 + D_z'^2) \, dS'. \tag{22}$$

§ 7. Our second special case is that of a particle having an electric moment, i.e. a small space S, with a total charge $\int \varrho \, dS = 0$, but with such a distribution of density that the integrals $\int \varrho x \, dS$, $\int \varrho y \, dS$, $\int \varrho z \, dS$ have values differing from 0. Let ξ, μ, ζ be the co-ordinates, taken relatively to a fixed point A of the particle, which may be called its centre, and let the electric moment be defined as a vector **P** whose components are

$$P_x = \int \varrho \xi \, dS, \quad P_y = \int \varrho \eta \, dS, \quad P_z = \int \varrho \zeta \, dS. \tag{23}$$

Then

$$\frac{dP_x}{dt} = \int \varrho u_x \, dS, \quad \frac{dP_y}{dt} = \int \varrho u_y \, dS, \quad \frac{dP_z}{dt} = \int \varrho u_z \, dS. \tag{24}$$

Of course, if ξ, η, ζ are treated as infinitely small, u_x, u_y, u_z must be so likewise. We shall neglect squares and products of these six quantities.

We shall now apply the equation (17) to the determination of the scalar potential ϕ' for an exterior point $P(x, y, z)$, at a finite distance from the polarized particle, and for the instant at which the local time of this point has some definite value t'. In doing so, we shall give the symbol $[\varrho]$, which, in (17), relates to the instant at which the local time in dS is $t' - r'/c$, a slightly different meaning. Distinguishing by r'_0 the value of r' for the centre A, we shall understand by $[\varrho]$ the value of the density existing in the element dS at the point (ξ, η, ζ), at the instant t_0 at which the local time of A is $t' - r_0/c$.

It may be seen from (5) that this instant precedes that for which we have to take the numerator in (17) by

$$\beta^2 \frac{v\xi}{c^2} + \frac{\beta(r'_0 - r')}{lc} = \beta^2 \frac{v\xi}{c^2} + \frac{\beta}{lc} \left(\xi \frac{\partial r'}{\partial x} + \eta \frac{\partial r'}{\partial y} + \zeta \frac{\partial r'}{\partial z} \right)$$

units of time. In this last expression we may put for the differential coefficients their values at the point A.

In (17) we have now to replace $[\varrho]$ by

$$[\varrho] + \beta^2 \frac{v\xi}{c^2} \left[\frac{\partial \varrho}{\partial t}\right] + \frac{\beta}{lc}\left(\xi \frac{\partial r'}{\partial x} + \eta \frac{\partial r'}{\partial y} + \zeta \frac{\partial r'}{\partial z}\right)\left[\frac{\partial \varrho}{\partial t}\right] \quad (25)$$

where $[\partial \varrho/\partial t]$ relates again to the time t_0. Now, the value of t' for which the calculations are to be performed having been chosen, this time t_0 will be a function of the co-ordinates x, y, z of the exterior point P. The value of $[\varrho]$ will therefore depend on these co-ordinates in such a way that

$$\frac{\partial [\varrho]}{\partial x} = -\frac{\beta}{lc} \frac{\partial r'}{\partial x}\left[\frac{\partial \varrho}{\partial t}\right], \text{ etc.}$$

by which (25) becomes

$$[\varrho] + \beta^2 \frac{v\xi}{c^2}\left[\frac{\partial \varrho}{\partial t}\right] - \left(\xi \frac{\partial [\varrho]}{\partial x} + \eta \frac{\partial [\varrho]}{\partial y} + \zeta \frac{\partial [\varrho]}{\partial z}\right).$$

Again, if henceforth we understand by r' what has above been called r'_0, the factor $1/r'$ must be replaced by

$$\frac{1}{r'} - \xi \frac{\partial}{\partial x}\left(\frac{1}{r'}\right) - \eta \frac{\partial}{\partial y}\left(\frac{1}{r'}\right) - \zeta \frac{\partial}{\partial z}\left(\frac{1}{r'}\right),$$

so that after all, in the integral (17), the element dS is multiplied by

$$\frac{[\varrho]}{r'} + \frac{\beta^2 v\xi}{c^2 r'}\left[\frac{\partial \varrho}{\partial t}\right] - \frac{\partial}{\partial x}\frac{\xi[\varrho]}{r'} - \frac{\partial}{\partial y}\frac{\eta[\varrho]}{r'} - \frac{\partial}{\partial z}\frac{\zeta[\varrho]}{r'}.$$

This is simpler than the primitive form, because neither r', nor the time for which the quantities enclosed in brackets are to be taken, depend on x, y, z. Using (23) and remembering that $\int \varrho \, dS = 0$, we get

$$\phi' = \frac{\beta^2 v}{4\pi c^2 r'}\left[\frac{\partial P_x}{\partial t}\right] - \frac{1}{4\pi}\left\{\frac{\partial}{\partial x}\frac{[P_x]}{r'} + \frac{\partial}{\partial y}\frac{[P_y]}{r'} + \frac{\partial}{\partial z}\frac{[P_z]}{r'}\right\},$$

a formula in which all the enclosed quantities are to be taken for the instant at which the local time of the centre of the particle is $t' - r'/c$.

We shall conclude these calculations by introducing a new vector $\mathbf{P}'$, whose components are

$$P'_x = \beta l P_x, \quad P'_y = l P_y, \quad P'_z = l P_z, \tag{26}$$

passing at the same time to x', y', z', t' as independent variables. The final result is

$$\phi' = \frac{v}{4\pi c^2 r'} \frac{\partial [P'_x]}{\partial t'} - \frac{1}{4\pi} \left\{ \frac{\partial}{\partial x'} \frac{[P'_x]}{r'} + \frac{\partial}{\partial y'} \frac{[P'_y]}{r'} + \frac{\partial}{\partial z'} \frac{[P'_z]}{r'} \right\}.$$

As to the formula (18) for the vector potential, its transformation is less complicated, because it contains the infinitely small vector $\mathbf{u}'$. Having regard to (8), (24), (26), and (5), I find

$$\mathbf{A}' = \frac{1}{4\pi c r'} \frac{\partial [\mathbf{P}']}{\partial t'}.$$

The field produced by the polarized particle is now wholly determined. The formula (13) leads to

$$\mathbf{D}' = -\frac{1}{4\pi c^2} \frac{\partial^2}{\partial t'^2} \frac{[\mathbf{P}']}{r'} + \frac{1}{4\pi} \operatorname{grad}' \left\{ \frac{\partial}{\partial x'} \cdot \frac{[P'_x]}{r'} + \frac{\partial}{\partial y'} \frac{[P'_y]}{r'} + \frac{\partial}{\partial z'} \frac{[P'_z]}{r'} \right\} \tag{27}$$

and the vector $\mathbf{H}'$ is given by (14). We may further use the equations (20), instead of the original formulae (10), if we wish to consider the forces exerted by the polarized particle on a similar one placed at some distance. Indeed, in the second particle, as well as in the first, the velocities $\mathbf{u}$ may be held to be infinitely small.

It is to be remarked that the formulae for a system without translation are implied in what precedes. For such a system the quantities with accents become identical to the corresponding ones without accents; also $\beta = 1$ and $l = 1$. The components of (27) are at the same time those of the electric force which is exerted by one polarized particle on another.

§ 8. Thus far we have used only the fundamental equations without any new assumptions. I shall now suppose *that the electrons, which I take to be spheres of radius R in the state of rest, have their dimensions changed by the effect of a translation, the dimensions in the direction of motion becoming βl times and those in perpendicular directions l times smaller*.

In this deformation, which may be represented by

$$\left(\frac{1}{\beta l}, \frac{1}{l}, \frac{1}{l}\right),$$

each element of colume is understood to preserve its charge.

Our assumption amounts to saying that in an electrostatic system Σ, moving with a velocity v, all electrons are flattened ellipsoids with their smaller axes in the direction of motion. If now, in order to apply the theorem of § 6, we subject the system to the deformation $(\beta l, l, l)$, we shall have again spherical electrons of radius R. Hence, if we alter the relative position of the centres of the electrons in Σ by applying the deformation $(\beta l, l, l)$, and if, in the points thus obtained, we place the centres of electrons that remain at rest, we shall get a system, identical to the imaginary system Σ', of which we have spoken in § 6. The forces in this system and those in Σ will bear to each other the relation expressed by (21).

In the second place I shall suppose *that the forces between uncharged particles, as well as those between such particles and electrons, are influenced by a translation in quite the same way as the electric forces in an electrostatic system*. In other terms, whatever be the nature of the particles composing a ponderable body, so long as they do not move relatively to each other, we shall have between the forces acting in a system (Σ') without, and the same system (Σ) with a translation, the relation specified in (21), if, as regards the relative position of the particles, Σ' is got from Σ by the deformation $(\beta l, l, l)$, or Σ from Σ' by the deformation

$$\left(\frac{1}{\beta l}, \frac{1}{l}, \frac{1}{l}\right).$$

We see by this that, as soon as the resulting force is zero for a particle in Σ', the same must be true for the corresponding particle in Σ. Consequently, if, neglecting the effects of molecular motion, we suppose each particle of a solid body to be in equilibrium under the action of the attractions and repulsions exerted by its neighbours, and if we take for granted that there is but one configuration of equilibrium, we may draw the conclusion that the system Σ', if the velocity v is imparted to it, will *of itself* change into the system Σ. In other terms, the translation will *produce* the deformation

$$\left(\frac{1}{\beta l}, \frac{1}{l}, \frac{1}{l}\right).$$

The case of molecular motion will be considered in § 12.

It will easily be seen that the hypothesis which was formerly advanced in connexion with Michelson's experiment, is implied in what has now been said. However, the present hypothesis is more general, because the only limitation imposed on the motion is that its velocity be less than that of light.

§ 9. We are now in a position to calculate the electromagnetic momentum of a single electron. For simplicity's sake I shall suppose the charge e to be uniformly distributed over the surface, so long as the electron remains at rest. Then a distribution of the same kind will exist in the system Σ' with which we are concerned in the last integral of (22). Hence

$$\int (D_y'^2 + D_z'^2)\, dS' = \frac{2}{3} \int D'^2\, dS' = \frac{e^2}{6\pi} \int_R^\infty \frac{dr}{r^2} = \frac{e^2}{6\pi R_i},$$

and

$$G_x = \frac{e^2}{6\pi c^2 R}\, \beta l v.$$

It must be observed that the product βl is a function of v and that, for reasons of symmetry, the vector **G** has the direction of the translation. In general, representing by **v** the velocity of this motion, we have the vector equation

$$\mathbf{G} = \frac{e^2}{6\pi c^2 R} \beta l \mathbf{v}. \tag{28}$$

Now, every change in the motion of a system will entail a corresponding change in the electromagnetic momentum and will therefore require a certain force, which is given in direction and magnitude by

$$\mathbf{F} = \frac{d\mathbf{G}}{dt}. \tag{29}$$

Strictly speaking, the formula (28) may only be applied in the case of a uniform rectilinear translation. On account of this circumstance—though (29) is always true—the theory of rapidly varying motions of an electron becomes very complicated, the more so, because the hypothesis of § 8 would imply that the direction and amount of the deformation are continually changing. It is, indeed, hardly probable that the form of the electron will be determined solely by the velocity existing at the moment considered.

Nevertheless, provided the changes in the state of motion be sufficiently slow, we shall get a satisfactory approximation by using (28) at every instant. The application of (29) to such a *quasi-stationary* translation, as it has been called by Abraham,* is a very simple matter. Let, at a certain instant, $\mathbf{a}_1$ be the acceleration in the direction of the path, and $\mathbf{a}_2$ the acceleration perpendicular to it. Then the force **F** will consist of two components, having the directions of these accelerations and which are given by

$$\mathbf{F}_1 = m_1 \mathbf{a}_1 \quad \text{and} \quad \mathbf{F}_2 = m_2 \mathbf{a}_2,$$

---

* Abraham, *Wied. Ann.*, **10**, 1903, p. 105.

if

$$m_1 = \frac{e^2}{6\pi c^2 R} \frac{d(\beta l v)}{dv} \quad \text{and} \quad m_2 = \frac{e^2}{6\pi c^2 R} \beta l. \tag{30}$$

Hence, in phenomena in which there is an acceleration in the direction of motion, the electron behaves as if it had a mass m_1; in those in which the acceleration is normal to the path, as if the mass were m_2. These quantities m_1 and m_2 may therefore properly be called the "longitudinal" and "transverse" electromagnetic masses of the electron. I shall suppose *that there is no other, no "true" or "material" mass*.

Since β and l differ from unity by quantities of the order v^2/c^2, we find for very small velocities

$$m_1 = m_2 = \frac{e^2}{6\pi c^2 R}.$$

This is the mass with which we are concerned, if there are small vibratory motions of the electrons in a system without translation. If, on the contrary, motions of this kind are going on in a body moving with the velocity v in the direction of the axis of x, we shall have to reckon with the mass m_1, as given by (30), if we consider the vibrations parallel to that axis, and with the mass m_2, if we treat of those that are parallel to OY or OZ. Therefore, in short terms, referring by the index Σ to a moving system and by Σ' to one that remains at rest,

$$m(\Sigma) = \left(\frac{d(\beta l v)}{dv}, \beta l, \beta l\right) m(\Sigma'). \tag{31}$$

§ 10. We can now proceed to examine the influence of the Earth's motion on optical phenomena in a system of transparent bodies. In discussing this problem we shall fix our attention on the variable electric moments in the particles or "atoms" of the system. To these moments we may apply what has been said in § 7. For the sake of simplicity we shall suppose that, in each particle, the charge is concentrated in a certain number of separate electrons, and that the "elastic" forces that act on one of these, and, conjointly with the electric forces, determine its motion, have their origin within the bounds of the *same* atom.

I shall show that, if we start from any given state of motion in a system without translation, we may deduce from it a corresponding state that can exist in the same system after a translation has been imparted to it, the kind of correspondence being as specified in what follows.

(*a*) Let A'_1, A'_2, A'_3, etc., be the centres of the particles in the system without translation (Σ'); neglecting molecular motions we shall assume these points to remain at rest. The system of points A_1, A_2, A_3, etc., formed by the centres of the particles in the moving system Σ, is obtained from A'_1, A'_2, A'_3, etc., by means of a deformation

$$\left(\frac{1}{\beta l}, \frac{1}{l}, \frac{1}{l}\right).$$

According to what has been said in § 8, the centres will of themselves take these positions A'_1, A'_2, A'_3, etc., if originally, before there was a translation, they occupied the positions A_1, A_2, A_3, etc.

We may conceive any point P' in the space of the system Σ' to be displaced by the above deformation, so that a definite point P of Σ corresponds to it. For two corresponding points P' and P we shall define corresponding instants, the one belonging to P', the other to P, by stating that the true time at the first instant is equal to the local time, as determined by (5) for the point P, at the second instant. By corresponding times for two corresponding *particles* we shall understand times that may be said to correspond, if we fix our attention on the *centres* A' and A of these particles.

(*b*) As regards the interior state of the atoms, we shall assume that the configuration of a particle A in Σ at a certain time may be derived by means of the deformation

$$\left(\frac{1}{\beta l}, \frac{1}{l}, \frac{1}{l}\right)$$

from the configuration of the corresponding particle in Σ', such as it is at the corresponding instant. In so far as this assumption relates to the form of the electrons themselves, it is implied in the first hypothesis of § 8.

Obviously, if we start from a state really existing in the system Σ', we have now completely defined a state of the moving system Σ. The question remains, however, whether this state will likewise be a possible one.

In order to judge of this, we may remark in the first place that the electric moments which we have supposed to exist in the moving system and which we shall denote by **P**, will be certain definite functions of the co-ordinates x, y, z of the centres A of the particles, or, as we shall say, of the co-ordinates of the particles themselves, and of the time t. The equations which express the relations between **P** on one hand and x, y, z, t on the other, may be replaced by other equations containing the vectors **P**' defined by (26) and the quantities x', y', z', t' defined by (4) and (5). Now, by the above assumptions a and b, if in a particle A of the moving system, whose co-ordinates are x, y, z, we find an electric moment **P** at the time t, or at the local time t', the vector **P**' given by (26) will be the moment which exists in the other system at the true time t' in a particle whose co-ordinates are x', y', z'. It appears in this way that the equations between **P**', x', y', z', t' are the same for both systems, the difference being only this, that for the system Σ' without translation these symbols indicate the moment, the co-ordinates, and the true time, whereas their meaning is different for the moving system, **P**', x', y', z', t' being here related to the moment **P**, the co-ordinates x, y, z and the general time t in the manner expressed by (26), (4), and (5).

It has already been stated that the equation (27) applies to both systems. The vector **D**' will therefore be the same in Σ' and Σ, provided we always compare corresponding places and times. However, this vector has not the same meaning in the two cases. In Σ' it represents the electric force, in Σ it is related to this force in the way expressed by (20). We may therefore conclude that the ponderomotive forces acting, in Σ and in Σ', on corresponding particles at corresponding instants, bear to each other the relation determined by (21). In virtue of our assumption (*b*), taken in connexion with the second hypothesis of § 8, the same relation will exist between the "elastic" forces; consequently, the formula (21) may also be regarded as indicating the relation between the total forces, acting on corresponding electrons, at corresponding instants.

It is clear that the state we have supposed to exist in the moving system will really be possible if, in Σ and Σ', the products of the mass *m* and the acceleration of an electron are to each other in the same relation as the forces, i.e. if

$$m\mathbf{a}(\Sigma) = \left(l^2, \frac{l^2}{\beta}, \frac{l^2}{\beta}\right) m\mathbf{a}(\Sigma'). \tag{32}$$

Now, we have for the accelerations

$$\mathbf{a}(\Sigma) = \left(\frac{l}{\beta^3}, \frac{l}{\beta^2}, \frac{l}{\beta^2}\right) \mathbf{a}(\Sigma') \tag{33}$$

as may be deduced from (4) and (5), and combining this with (32), we find for the masses

$$m(\Sigma) = (\beta^3 l, \beta l, \beta l) m(\Sigma').$$

If this is compared with (31), it appears that, whatever be the value of *l*, the condition is always satisfied, as regards the masses with which we have to reckon when we consider vibrations perpendicular to the translation. The only condition we have to impose on *l* is therefore

$$\frac{d(\beta l v)}{dv} = \beta^3 l.$$

But, on account of (3),

$$\frac{d(\beta v)}{dv} = \beta^3,$$

138 SPECIAL RELATIVITY

so that we must put
$$\frac{dl}{dv} = 0, \quad l = \text{const.}$$

The value of the constant must be unity, because we know already that, for $v = 0, l = 1$.

We are therefore led to suppose *that the influence of a translation on the dimensions (of the separate electrons and of a ponderable body as a whole) is confined to those that have the direction of the motion, these becoming β times smaller than they are in the state of rest*. If this hypothesis is added to those we have already made, we may be sure that two states, the one in the moving system, the other in the same system while at rest, corresponding as stated above, may both be possible. Moreover, this correspondence is not limited to the electric moments of the particles. In corresponding points that are situated either in the ether between the particles, or in that surrounding the ponderable bodies, we shall find at corresponding times the same vector **D**' and, as is easily shown, the same vector **H**'. We may sum up by saying: If, in the system without translation, there is a state of motion in which, at a definite place, the components of **P**, **D**, and **H** are certain functions of the time, then the same system after it has been put in motion (and thereby deformed) can be the seat of a state of motion in which, at the corresponding place, the components of **P**', **D**', and **H**' are the same functions of the local time.

There is one point which requires further consideration. The values of the masses m_1 and m_2 having been deduced from the theory of quasi-stationary motion, the question arises, whether we are justified in reckoning with them in the case of the rapid vibrations of light. Now it is found on closer examination that the motion of an electron may be treated as quasi-stationary if it changes very little during the time a light-wave takes to travel over a distance equal to the diameter. This condition is fulfilled in optical phenomena, because the diameter of an electron is extremely small in comparison with the wave-length.

§ 11. It is easily seen that the proposed theory can account for a large number of facts.

Let us take in the first place the case of a system without translation, in some parts of which we have continually $\mathbf{P} = 0$, $\mathbf{D} = 0$, $\mathbf{H} = 0$. Then, in the corresponding state for the moving system, we shall have in corresponding parts (or, as we may say, in the same parts of the deformed system) $\mathbf{P}' = 0$, $\mathbf{D}' = 0$, $\mathbf{H}' = 0$. These equations implying $\mathbf{P} = 0$, $\mathbf{D} = 0$, $\mathbf{H} = 0$, as is seen by (26) and (6), it appears that those parts which are dark while the system is at rest, will remain so after it has been put in motion. It will therefore be impossible to detect an influence of the Earth's motion on any optical experiment, made with a terrestrial source of light, in which the geometrical distribution of light and darkness is observed. Many experiments on interference and diffraction belong to this class.

In the second place, if, in two points of a system, rays of light of the same state of polarization are propagated in the same direction, the ratio between the amplitudes in these points may be shown not to be altered by a translation. The latter remark applies to those experiments in which the intensities in adjacent parts of the field of view are compared.

The above conclusions confirm the results which I formerly obtained by a similar train of reasoning, in which, however, the terms of the second order were neglected. They also contain an explanation of Michelson's negative result, more general than the one previously given, and of a somewhat different form; and they show why Rayleigh and Brace could find no signs of double refraction produced by the motion of the Earth.

As to the experiments of Trouton and Noble, their negative result becomes at once clear, if we admit the hypotheses of § 8. It may be inferred from these and from our last assumption (§ 10) that the only effect of the translation must have been a contraction of the whole system of electrons and other particles constituting the charged condenser and the beam and thread of the torsion-balance. Such a contraction does not give rise to a sensible change of direction.

It need hardly be said that the present theory is put forward with all due reserve. Though it seems to me that it can account for all well-established facts, it leads to some consequences that cannot as yet be put to the test of experiment. One of these is that the result of Michelson's experiment must remain negative, if the interfering rays of light are made to travel through some ponderable transparent body.

Our assumption about the contraction of the electrons cannot in itself be pronounced to be either plausible or inadmissible. What we know about the nature of electrons is very little, and the only means of pushing our way farther will be to test such hypotheses as I have here made. Of course, there will be difficulties, e.g. as soon as we come to consider the rotation of electrons. Perhaps we shall have to suppose that in those phenomena in which, if there is no translation, spherical electrons rotate about a diameter, the points of the electrons in the moving system will describe elliptic paths, corresponding, in the manner specified in § 10, to the circular paths described in the other case.

§ 12. There remain to be said a few words about molecular motion. We may conceive that bodies in which this has a sensible influence or even predominates, undergo the same deformation as the systems of particles of constant relative position of which alone we have spoken till now. Indeed, in two systems of molecules Σ' and Σ, the first without and the second with a translation, we may imagine molecular motions corresponding to each other in such a way that, if a particle in Σ' has a certain position at a definite instant, a particle in Σ occupies at the corresponding instant the corresponding position. This being assumed, we may use the relation (33) between the accelerations in all those cases in which the velocity of molecular motion is very small as compared with v. In these cases the molecular forces may be taken to be determined by the relative positions, independently of the velocities of molecular motion. If, finally, we suppose these forces to be limited to such small distances that, for particles acting on each other, the difference of local times may be neglected, one of the particles, together with those which lie in its sphere of attraction or repulsion, will form a system which undergoes the often mentioned deformation. In virtue of the second hypothesis of § 8 we may therefore apply to the resulting molecular force acting on a particle, the equation (21). Consequently, the proper relation between the forces and the accelerations will exist in the two cases, if we suppose *that the masses of all particles are influenced by a translation to the same degree as the electromagnetic masses of the electrons.*

§ 13. The values (30), which I have found for the longitudinal and transverse masses of an electron, expressed in terms of its velocity, are not the same as those that had been previously obtained by Abraham. The ground for this difference is to be sought solely in the circumstance that, in his theory, the electrons are treated as spheres of invariable dimensions. Now, as regards the transverse mass, the results of Abraham have been confirmed in a most remarkable way by Kaufmann's measurements of the deflexion of radium-rays in electric and magnetic fields. Therefore, if there is not to be a most serious objection to the theory I have now proposed, it must be possible to show that those measurements agree with my values nearly as well as with those of Abraham.

I shall begin by discussing two of the series of measurements published by Kaufmann[*] in 1902. From each series he has deduced two quantities η and ζ, the "reduced" electric and magnetic deflexions, which are related as follows to the ratio $\gamma = v/c$:

$$\gamma = k_1 \frac{\zeta}{\eta}, \quad \psi(\gamma) = \frac{\eta}{k_2 \zeta^2}. \tag{34}$$

---

[*] Kaufmann, *Physik. Zeitschr.*, **4**, 1902, p. 55.

Here $\psi(\gamma)$ is such a function, that the transverse mass is given by

$$m_2 = \frac{3}{4} \cdot \frac{e^2}{6\pi c^2 R} \psi(\gamma), \tag{35}$$

whereas k_1 and k_2 are constant in each series.

It appears from the second of the formulae (30) that my theory leads likewise to an equation of the form (35); only Abraham's function $\psi(\gamma)$ must be replaced by

$$\tfrac{4}{3}\beta = \tfrac{4}{3}(1-\gamma^2)^{-1/2}.$$

Hence, my theory requires that, if we substitute this value for $\psi(\gamma)$ in (34), these equations shall still hold. Of course, in seeking to obtain a good agreement, we shall be justified in giving to k_1 and k_2 other values than those of Kaufmann, and in taking for every measurement a proper value of the velocity v, or of the ratio γ. Writing sk_1, $\tfrac{3}{4}k_2'$ and γ' for the new values, we may put (34) in the form

$$\gamma' = sk_1 \frac{\zeta}{\eta}: \tag{36}$$

and

$$(1-\gamma'^2)^{-1/2} = \frac{\eta}{k_2'\zeta^2}. \tag{37}$$

Kaufmann has tested his equations by choosing for k_1 such a value that, calculating γ and k_2 by means of (34), he obtained values for this latter number which, as well as might be, remained constant in each series. This constancy was the proof of a sufficient agreement.

I have followed a similar method, using, however, some of the numbers calculated by Kaufmann. I have computed for each measurement the value of the expression

$$k_2' = (1-\gamma'^2)^{1/2} \psi(\gamma) k_2, \tag{38}$$

that may be got from (37) combined with the second of the equations (34). The values of $\psi(\gamma)$ and k_2 have been taken from Kaufmann's tables, and for γ' I have substituted the value he has found for γ, multiplied by s, the latter coefficient being chosen with a view to obtaining a good constancy of (38). The results are contained in the tables below, corresponding to the Tables III and IV in Kaufmann's paper.

The constancy of k_2' is seen to come out no less satisfactorily than that of k_2, the more so as in each case the value of s has been determined by means of only two measurements. The coefficient has been so chosen that for these two observations, which were in Table III the first and the last but one, and in Table IV the first and the last, the values of k_2' should be proportional to those of k_2.

III. $s = 0.933$.

γ.	$\psi(\gamma)$.	k_2.	γ'.	k_2'.
0·851	2·147	1·721	0·794	2·246
0·766	1·86	1·736	0·715	2·258
0·727	1·78	1·725	0·678	2·256
0·6615	1·66	1·727	0·617	2·256
0·6075	1·595	1·655	0·567	2·175

IV. $s = 0.954$.

γ.	$\psi(\gamma)$.	k_2.	γ'.	k_2'.
0·963	3·28	8·12	0·919	10·36
0·949	2·86	7·99	0·905	9·70
0·933	2·73	7·46	0·890	9·28
0·883	2·31	8·32	0·842	10·36
0·860	2·195	8·09	0·820	10·15
0·830	2·06	8·13	0·792	10·23
0·801	1·96	8·13	0·764	10·28
0·777	1·89	8·04	0·741	10·20
0·752	1·83	8·02	0·717	10·22
0·732	1·785	7·97	0·698	10·18

NOTES ON EXTRACT 4

POINCARÉ starts by considering the experiments of Michelson and the explanation of Lorentz and Fitzgerald, and he has already in mind the later article of Lorentz (Extract 3). His own approach has been quite independent of Lorentz, and he remarks that the results which he has obtained agree in all important respects with those of Lorentz. The first section recapitulates the results already given by Lorentz, and he next proceeds, with typically French elegance, to deduce all of these results from a variational principle. This deduction is carried out in order that in the third section Poincaré can relate the invariance under the Lorentz transformation to the invariance of the variational principle. In the fourth section he goes on to show that the transformations do indeed form a group. Subsequent sections (which are omitted) are then concerned with rather technical matters of less interest now, but at the end of the paper there is a section in which Poincaré attempts to relate what he has found to the problem of gravitation. This problem is one which will occupy us exclusively in the succeeding book, since it found its complete solution not by means of field theories of the kind envisaged by Poincaré but by an entirely reformulated theory (general relativity) some years later It is, however, of the greatest interest to observe what an extremely sophisticated gravitational theory can be produced by Poincaré in a Lorentz invariant fashion.

One mathematical point needs to be noted. Poincaré adheres to the old usage of d for both partial and ordinary differentiation. He then introduces what would usually be a symbol of partial differentiation, with a specialised meaning, in Section 2.

EXTRACT 4[*]

The dynamics of the Electron

By H. POINCARÉ

Introduction

It would seem at first sight that the aberration of light and the optical and electrical effects related thereto should afford a means of determining the absolute motion of the Earth, or rather its motion relative to the ether instead of relative to the other celestial bodies. An attempt at this was made, indeed, by Fresnel, but he soon perceived that the Earth's motion does not affect the laws of refraction and reflection. Similar experiments, such as that using a waterfilled telescope, or any in which only the first-order terms relative to the aberration were considered, likewise yielded only negative results. The explanation of this was soon found; but Michelson, who devised an experiment wherein the terms involving the square of the aberration should be detectable, was equally unsuccessful.

This impossibility of experimentally demonstrating the absolute motion of the Earth appears to be a general law of Nature; it is reasonable to assume the existence of this law, which we shall call the *relativity postulate*, and to assume that it is universally valid. Whether this postulate, which so far is in agreement with experiment, be later confirmed or disproved by more accurate tests, it is, in any case, of interest to see what consequences follow from it.

[* *Rend. del Circ. Mat. di Palermo* **21**, 129–46 and 166–75 (1906).]

One explanation, suggested by Lorentz and Fitzgerald, involves the hypothesis that all bodies undergo a contraction in the direction of the Earth's motion, of an amount proportional to the square of the aberration; such a contraction, which we shall call the *Lorentz contraction*, would explain the result of Michelson's experiment and of all others conducted heretofore. The hypothesis would nevertheless be inadequate if the relativity postulate were valid in its most general form.

Lorentz has sought to extend and modify the hypothesis so as to make it fully compatible with the relativity postulate. This he has succeeded in doing, in his paper "Electromagnetic phenomena in a system moving with any velocity smaller than that of light" (*Proceedings of the Section of Sciences, Koninklijke Akademie van Wetenschappen te Amsterdam* **6**, 809–831, 1904).

In view of the importance of this problem, I resolved to examine it further. The results which I have obtained agree with those of Lorentz in all the principal points, and I have needed only to modify and augment them in certain details. These differences, which are of but minor importance, will be shown in later sections.

Lorentz's concept may be summarised thus: if a common translatory motion may be imparted to the entire system without any alteration of the observable phenomena, then the equations of an electromagnetic medium are unaltered by certain transformations, which we shall call *Lorentz transformations*. In this way two systems, of which one is fixed and the other is in translatory motion, become exact images of each other.

Langevin[†] sought to derive a modification of Lorentz's concept. Both authors consider that an electron in motion assumes the form of an oblate spheroid; but Lorentz considers that two of the axes of this spheroid remain constant, whereas Langevin supposes that its volume remains constant. These two authors

[†] Langevin had been anticipated by Bucherer of Bonn, who earlier put forward the same idea. See A. H. Bucherer, *Mathematische Einführung in die Elektronentheorie*, Teubner, Leipzig, 1904.

have shown that the two hypotheses are in agreement with the experiments of Kaufmann, as is Abraham's original hypothesis of a rigid spherical electron.

The advantage of Langevin's theory is that it involves only the electromagnetic forces and the constraints; but it is not compatible with the relativity postulate. This was shown by Lorentz, and I have likewise proved it by a different method, based upon the use of group theory.

We must return therefore to Lorentz's theory, but, in order to maintain this free from unacceptable contradictions, a special force must be invoked to account both for the contraction and for the constancy of two of the axes. I have attempted to determine this force, and have found that *it can be regarded as a constant external pressure acting upon an electron capable of deformation and compression, the work done being proportional to the change in the volume of the electron.*

Then, if the inertia of matter is exclusively of electromagnetic origin, as has been customarily supposed since Kaufmann's experiment, and if all forces (other than the constant pressure to which I have just alluded) are of electromagnetic origin, the relativity postulate can be accepted as strictly valid. I show this by means of a very simple calculation based upon the principle of least action.

But this is not all. Lorentz, in his paper already mentioned, has deemed it necessary to extend his hypothesis in such a manner that the postulate remains valid when there exist forces other than the electromagnetic forces. In Lorentz's view, all forces, no matter how originating, are affected by the Lorentz transformation (and therefore by a translatory motion) in the same manner as the electromagnetic forces.

It was necessary to consider this hypothesis more closely, and in particular to ascertain the changes which it would compel us to apply to the laws of gravitation.

First of all, we find that gravitational action would be propagated with the velocity of light, and not instantaneously. This might in itself appear to be sufficient reason to reject the hypothesis, for Laplace has shown that such propagation cannot occur. But, in fact, the effects of this are largely counterbalanced by another phenomenon, and there is, therefore, no contradiction between the proposed law and astronomical observations.

The question arises whether it is possible to discover a law which satisfies Lorentz's condition and which yet reduces to Newton's law whenever the velocities of the bodies are so small that the squares of these velocities (and the products of the accelerations and the distances) may be neglected in comparison with the square of the velocity of light.

It will be seen later than the answer must be affirmative.

Is the law, thus modified, compatible with astronomical observations?

At first sight it appears to be so, but a more detailed discussion is necessary to settle the question.

Even assuming, however, that the new hypothesis survives this test, what conclusion is to be drawn? If the gravitational attraction is propagated with the velocity of light, this cannot occur by mere chance, but must be dependent on the ether; we should then have to investigate the nature of this dependence, and attempt to relate it to other such dependences.

We cannot be satisfied with formulae that are merely placed side by side and agree only by a lucky chance; these formulae must, as it were, interlock. The mind will consent only when it sees the reason for the agreement, and when this agreement even seems to have been predictable.

But the matter may be viewed in a different light, as an analogy will show. Let us imagine some astronomer before Copernicus, pondering upon the Ptolemaic system. He would notice that, for every planet, either the epicycle or the deferent is traversed in the same time. This cannot be due to chance, and there must be some mysterious bond between all the planets of the system.

Then Copernicus, by a simple change of the co-ordinate axes which were supposed fixed, did away with this seeming relationship: every planet described one circular orbit only, and the periods of revolution became independent of one another—until Kepler once more established the relationship that had apparently been destroyed.

Now, there may be an analogy with our problem. If we assume the relativity postulate, we find a quantity common to the law of gravitation and the laws of electromagnetism, and this quantity is the velocity of light; and this same quantity appears in every other force, of whatever origin. There can be only two explanations.

Either, everything in the universe is of electromagnetic origin; or, this constituent which appears common to all the phenomena of physics has no real existence, but arises from our methods of measurement. What are these methods? One might first reply, the bringing into juxtaposition of objects regarded as invariable solid things; but this is no longer so in our present theory, if the Lorentz contraction is assumed. In this theory, two lengths are by definition equal if they are traversed by light in the same time.

Perhaps the abandonment of this definition would suffice to overthrow Lorentz's theory as decisively as the system of Ptolemy was by the work of Copernicus. Should this ever happen, it would by no means argue the futility of Lorentz's analysis: whatever the faults of the Ptolemaic theory, it was the necessary foundation for Copernicus to build upon.

I have therefore not hesitated to publish these incomplete results, even though at the present time the entire theory may seem to be threatened by the discovery of cathode rays.

§ 1. The Lorentz Transformation

Lorentz has adopted a particular system of units, such that the factors of 4π no longer appear in the formulae. I shall do likewise, and moreover I shall choose the units of length and of time in such a way that the velocity of light is equal to unity. Then, if f, g, h denote the electrical displacement; α, β, γ the magnetic force; F, G, H the vector potential; ψ the scalar potential; ϱ the electrical charge density; ξ, η, ζ the velocity of the electron; u, v, w the current, the fundamental equations become

$$\left.\begin{aligned}
u &= \frac{df}{dt}+\varrho\xi = \frac{d\gamma}{dy}-\frac{d\beta}{dz}, \\
\alpha &= \frac{dH}{dy}-\frac{dG}{dz}, \quad f = -\frac{dF}{dt}-\frac{d\psi}{dx}, \\
\frac{d\alpha}{dt} &= \frac{dg}{dz}-\frac{dh}{dy}, \quad \frac{d\varrho}{dt}+\Sigma\frac{d(\varrho\xi)}{dx} = 0, \\
\Sigma\frac{df}{dx} &= \varrho, \quad \frac{d\psi}{dt}+\Sigma\frac{dF}{dx} = 0, \\
\Box &= \varDelta-\frac{d^2}{dt^2} = \Sigma\frac{d^2}{dx^2}-\frac{d^2}{dt^2}, \\
\Box\psi &= -\varrho, \quad \Box F = -\varrho\xi.
\end{aligned}\right\} \quad (1)$$

An elementary particle of matter, having a volume $dx\,dy\,dz$, is acted upon by a mechanical force, whose components $X\,dx\,dy\,dz$, $Y\,dx\,dy\,dz$, $Z\,dx\,dy\,dz$ are given by the formula

$$X = \varrho f+\varrho(\eta\gamma-\zeta\beta). \qquad (2)$$

These equations can be subjected to a remarkable transformation discovered by Lorentz, the significance of which is that it explains why no experimental demonstration of the absolute motion of the universe is possible. If we put

$$x' = kl(x+\varepsilon t), \quad t' = kl(t+\varepsilon x), \quad y' = ly, \quad z' = lz, \qquad (3)$$

where l and ε are any constants, and

$$k = \frac{1}{\sqrt{(1-\varepsilon^2)}},$$

and if we also put

$$\square' = \Sigma \frac{d^2}{dx'^2} - \frac{d^2}{dt'^2},$$

then

$$\square' = \square \cdot l^{-2}.$$

Let a sphere be carried along with the electron in a uniform translatory motion, and let the equation of this moving sphere be

$$(x-\xi t)^2 + (y-\eta t)^2 + (z-\zeta t)^2 = r^2;$$

the volume of the sphere is then $\frac{4}{3}\pi r^3$.

The foregoing transformation will change the sphere into an ellipsoid, whose equation is easily found. From the equations (3), it immediately follows that

$$x = \frac{k}{l}(x' - \varepsilon t'), \quad t = \frac{k}{l}(t' - \varepsilon x'), \quad y = \frac{y'}{l}, \quad z = \frac{z'}{l}. \quad (3')$$

The equation of the ellipsoid is then

$$k^2(x' - \varepsilon t' - \xi t' + \varepsilon \xi x')^2 + (y' - \eta k t' + \eta k \varepsilon x')^2$$
$$+ (z' - \zeta k t' + \zeta k \varepsilon x')^2 = l^2 r^2.$$

The ellipsoid moves uniformly; when $t' = 0$, it is

$$k^2 x'^2 (1+\xi\varepsilon)^2 + (y' + \eta k \varepsilon x')^2 + (z' + \zeta k \varepsilon x')^2 = l^2 r^2,$$

and its volume is

$$\frac{4}{3}\pi r^3 \frac{l^3}{k(1+\xi\varepsilon)}.$$

If the charge on an electron is to be unaltered by the transformation, and if the new electrical charge density be denoted by

ϱ', it follows that
$$\varrho' = \frac{k}{l^3}(\varrho + \varepsilon\varrho\xi). \tag{4}$$

The new velocities ξ', η', ζ' will be given by

$$\xi' = \frac{dx'}{dt'} = \frac{d(x+\varepsilon t)}{d(t+\varepsilon x)} = \frac{\xi+\varepsilon}{1+\varepsilon\xi},$$

$$\eta' = \frac{dy'}{dt'} = \frac{dy}{k\,d(t+\varepsilon x)} = \frac{\eta}{k(1+\varepsilon\xi)},$$

$$\zeta' = \frac{\zeta}{k(1+\varepsilon\xi)},$$

whence

$$\varrho'\xi' = \frac{k}{l^3}(\varrho\xi+\varepsilon\varrho), \quad \varrho'\eta' = \frac{1}{l^3}\varrho\eta, \quad \varrho'\zeta' = \frac{1}{l^3}\varrho\zeta. \tag{4'}$$

Here I must for the first time indicate a disagreement with Lorentz's analysis. Lorentz (*op. cit.*, page 813, formulae (7) and (8)) writes, in our notation,

$$\varrho' = \frac{1}{kl^3}\varrho, \quad \xi' = k^2(\xi+\varepsilon), \quad \eta' = k\eta, \quad \zeta' = k\zeta.$$

These lead to the same relationships

$$\varrho'\xi' = \frac{k}{l^3}(\varrho\xi+\varepsilon\varrho), \quad \varrho'\eta' = \frac{1}{l^3}\varrho\eta, \quad \varrho'\zeta' = \frac{1}{l^3}\varrho\zeta,$$

but with a different value of ϱ'.

It should be noticed that formulae (4) and (4') satisfy the continuity condition

$$\frac{d\varrho'}{dt} + \Sigma \frac{d(\varrho'\xi')}{dx'} = 0.$$

For, let λ be an undetermined coefficient, and D the Jacobian of

$$t+\lambda\varrho, \quad x+\lambda\varrho\xi, \quad y+\lambda\varrho\eta, \quad z+\lambda\varrho\zeta \tag{5}$$

with respect to t, x, y, z. Then

$$D = D_0 + D_1\lambda + D_2\lambda^2 + D_3\lambda^3 + D_4\lambda^4,$$

with $\quad D_0 = 1, \quad D_1 = \dfrac{d\varrho}{dt} + \Sigma \dfrac{d(\varrho\xi)}{dx} = 0.$

Let $\lambda' = l^2\lambda$; then the four functions

$$t' + \lambda'\varrho', \quad x' + \lambda'\varrho'\xi', \quad y' + \lambda'\varrho'\eta', \quad z' + \lambda'\varrho'\zeta' \qquad (5')$$

are related to the functions (5) by the same linear relationships as those which exist between the old and new variables. If, therefore, D' denotes the Jacobian of the functions (5') with respect to the new variables, then

$$D' = D, \quad D' = D'_0 + D'_1\lambda' + \ldots + D'_4\lambda'^4,$$

whence

$$D'_0 = D_0 = 1, \quad D'_1 = l^{-2}D_1 = 0$$
$$= \frac{d\varrho'}{dt'} + \Sigma \frac{d(\varrho'\xi')}{dx'}, \qquad \text{q.e.d.}$$

With Lorentz's hypothesis, this condition would not be fulfilled, since the value of ϱ' is not the same.

The new vector and scalar potentials will be defined so as to satisfy the conditions

$$\Box'\psi' = -\varrho', \quad \Box'F' = -\varrho'\xi'. \qquad (6)$$

Hence we find

$$\psi' = \frac{k}{l}(\psi + \varepsilon F), \quad F' = \frac{k}{l}(F + \varepsilon\psi), \quad G' = \frac{1}{l}G, \quad H' = \frac{1}{l}H. \qquad (7)$$

These formulae are noticeably different from those of Lorentz, but the difference rests, ultimately, only on the definitions used.

The new electric and magnetic fields will be defined so as to satisfy the equations

$$f' = -\frac{dF'}{dt'} - \frac{d\psi'}{dx'}, \quad \alpha' = \frac{dH'}{dy'} - \frac{dG'}{dz'}. \qquad (8)$$

It is easily seen that

$$\frac{d}{dt'} = \frac{k}{l}\left(\frac{d}{dt} - \varepsilon\frac{d}{dx}\right), \quad \frac{d}{dx'} = \frac{k}{l}\left(\frac{d}{dx} - \varepsilon\frac{d}{dt}\right),$$

$$\frac{d}{dy'} = \frac{1}{l}\frac{d}{dy}, \quad \frac{d}{dz'} = \frac{1}{l}\frac{d}{dz},$$

and therefore

$$\left.\begin{array}{l}f' = \dfrac{1}{l^2}f, \quad g' = \dfrac{k}{l^2}(g+\varepsilon\gamma), \quad h' = \dfrac{k}{l^2}(h-\varepsilon\beta), \\[2mm] \alpha' = \dfrac{1}{l^2}\alpha, \quad \beta' = \dfrac{k}{l^2}(\beta-\varepsilon h), \quad \gamma' = \dfrac{k}{l^2}(\gamma+\varepsilon g).\end{array}\right\} \quad (9)$$

These formulae are identical with those of Lorentz.

Our transformation does not affect equations (1): the continuity condition and equations (6) and (8) are identical with some of the equations (1) if the primes are omitted.

The equations (6), together with the continuity condition, give

$$\frac{d\varphi'}{dt'} + \Sigma\frac{dF'}{dx'} = 0. \qquad (10)$$

We have only to prove that

$$\frac{df'}{dt'} + \varrho'\xi' = \frac{d\gamma'}{dy'} - \frac{d\beta'}{dz'}, \quad \frac{d\alpha'}{dt'} = \frac{dg'}{dz'} - \frac{dh'}{dy'},$$

$$\Sigma\frac{df'}{dx'} = \varrho',$$

and it is easily seen that these relationships necessarily follow from equations (6), (8) and (10).

Let us now make a comparison of the forces before and after the transformation.

Let X, Y, Z be the force before the transformation, and X', Y', Z' the force after it, both per unit volume. If X' is to satisfy the

same equations as before the transformation, we must have
$$X' = \varrho'f' + \varrho'(\eta'\gamma' - \zeta'\beta'),$$
$$Y' = \varrho'g' + \varrho'(\zeta'\alpha' - \xi'\gamma'),$$
$$Z' = \varrho'h' + \varrho'(\xi'\beta' - \eta'\alpha'),$$

or, substituting the expressions (4), (4′) and (9) and using equations (2),

$$\left.\begin{array}{l} X' = \dfrac{k}{l^5}(X + \varepsilon\Sigma X\xi), \\[6pt] Y' = \dfrac{1}{l^5} Y, \\[6pt] Z' = \dfrac{1}{l^5} Z. \end{array}\right\} \quad (11)$$

If X_1, Y_1, Z_1 denote the components of the force per unit electric charge on the electron, and X'_1, Y'_1, Z'_1 the same quantities after the transformation, then
$$\begin{array}{ll} X_1 = f + \eta\gamma - \zeta\beta, & X'_1 = f' + \eta'\gamma' - \zeta'\beta', \\ X = \varrho X_1, & X' = \varrho' X'_1, \end{array}$$
and we should obtain

$$\left.\begin{array}{l} X'_1 = \dfrac{k}{l^5}\dfrac{\varrho}{\varrho'}(X_1 + \varepsilon\Sigma X_1\xi), \\[6pt] Y'_1 = \dfrac{1}{l^5}\dfrac{\varrho}{\varrho'} Y_1, \\[6pt] Z'_1 = \dfrac{1}{l^5}\dfrac{\varrho}{\varrho'} Z_1. \end{array}\right\} \quad (11')$$

Lorentz's result was, in our notation (*op. cit.*, page 813, formula (10)),

$$\left.\begin{array}{l} X_1 = l^2 X'_1 - l^2\varepsilon(\eta'g' + \zeta'h'), \\[6pt] Y_1 = \dfrac{l^2}{k} Y'_1 + \dfrac{l^2\varepsilon}{k}\xi'g', \\[6pt] Z_1 = \dfrac{l^2}{k} Z'_1 + \dfrac{l^2\varepsilon}{k}\xi'h'. \end{array}\right\} \quad (11'')$$

Before proceeding, it is necessary to ascertain the reason for this considerable difference. It occurs, evidently, because the formulae for ξ', η', ζ' are not the same, whereas those for the electric and magnetic fields are the same.

If the inertia of the electrons is of purely electromagnetic origin, and if moreover they are subject only to forces of electromagnetic origin, the condition of equilibrium requires that, within the electrons,

$$X = Y = Z = 0.$$

From the relations (11), these are clearly equivalent to

$$X' = Y' = Z' = 0.$$

Thus the equilibrium conditions are unaffected by the transformation.

Unfortunately, such a simple hypothesis is inadmissible. For, if we assume that $\xi = \eta = \zeta = 0$, the conditions $X = Y = Z = 0$ will imply that $f = g = h = 0$, and therefore

$$\Sigma \frac{df}{dx} = 0, \quad \text{i.e. } \varrho = 0.$$

Similar results would be obtained in the general case. Hence we must assume that there are not only electromagnetic forces but also either other forces or constraints. We then have to determine the conditions governing these forces or constraints such that the equilibrium of the electrons is unaffected by the transformation. This will be done in a subsequent section.

§ 2. The Principle of Least Action

Lorentz's derivation of his equations from the principle of least action is well known. I shall, however, discuss this point further (although I have nothing essential to add to Lorentz's analysis), since I prefer to present it in a slightly different form, which will be of use later. I write

$$J = \int dt\, d\tau [\tfrac{1}{2}\Sigma f^2 + \tfrac{1}{2}\Sigma\alpha^2 - \Sigma Fu], \qquad (1)$$

with f, α, F, u, etc., assumed subject to the following conditions and those obtained from them by symmetry:

$$\Sigma \frac{df}{dx} = \varrho, \quad \alpha = \frac{dH}{dy} - \frac{dG}{dz}, \quad u = \frac{df}{dt} + \varrho\xi. \qquad (2)$$

The integral J is taken over the following ranges:

(a) the whole of space, for the volume element $d\tau = dx\, dy\, dz$;
(b) the interval between t_0 and t_1, for the time element dt.

According to the principle of least action, the integral J must have a minimum value when the quantities in it satisfy:

(a) the conditions (2);
(b) the condition that the system is in specified states at the limiting times t_0 and t_1.

The latter condition enables us to transform the integrals, using an integration by parts with respect to the time. For, given an integral of the form

$$\int dt\, d\tau\, A\, \frac{dB\, \delta C}{dt},$$

where C is one of the quantities defining the state of the system, and δC the variation of C, integration by parts with respect to the time shows that this integral is equal to

$$\int d\tau \left[AB\, \delta C \right]_{t=t_0}^{t=t_1} - \int dt\, d\tau\, \frac{dA}{dt}\, dB\, \delta C.$$

Since the state of the system at the limiting times is specified, $\delta C = 0$ for $t = t_0$ and for $t = t_1$; the first integral is therefore zero, and only the second integral remains.

We can effect a similar integration by parts with respect to x, y or z, since

$$\int A \frac{dB}{dx} dx\, dy\, dz\, dt = \int AB\, dy\, dz\, dt - \int B \frac{dA}{dx} dx\, dy\, dz\, dt.$$

The integrations extend to infinity, and in the first integral on the right-hand side we must therefore put $x = \pm \infty$; this integral is then zero, because all the functions are assumed to tend to zero at infinity, and we have

$$\int A \frac{dB}{dx} d\tau\, dt = -\int B \frac{dA}{dx} d\tau\, dt.$$

If the system were assumed subject to constraints, the constraint conditions would have to be included among the conditions to be satisfied by the various quantities appearing in the integral J.

First, let F, G, H receive increments δF, δG, δH; then

$$\delta \alpha = \frac{d\, \delta H}{dy} - \frac{d\, \delta G}{dz}.$$

We must have

$$\delta J = \int dt\, d\tau \left[\Sigma \alpha \left(\frac{d\, \delta H}{dy} - \frac{d\, \delta G}{dz} \right) - \Sigma u\, \delta F \right] = 0,$$

or, on integrating by parts,

$$\delta J = \int dt\, d\tau \left[\Sigma \left(\delta G \frac{\delta \alpha}{\delta z} - \delta H \frac{d\alpha}{dy} \right) - \Sigma u\, \delta F \right]$$

$$= -\int dt\, d\tau \Sigma\, \delta F \left(u - \frac{d\gamma}{dy} + \frac{d\beta}{dz} \right) = 0,$$

whence, equating to zero the coefficient of the arbitrary quantity δF,

$$u = \frac{d\gamma}{dy} - \frac{d\beta}{dz}. \tag{3}$$

From this we obtain (using an integration by parts)

$$\int \Sigma Fu \, d\tau = \int \Sigma F\left(\frac{d\gamma}{dy} - \frac{d\beta}{dz}\right) d\tau$$
$$= \int \Sigma\left(\beta \frac{dF}{dz} - \gamma \frac{dF}{dy}\right) d\tau$$
$$= \int \Sigma\alpha\left(\frac{dH}{dy} - \frac{dG}{dz}\right) d\tau,$$

or

$$\int \Sigma Fu \, d\tau = \int \Sigma\alpha^2 \, d\tau,$$

whence finally

$$J = \int dt \, d\tau \left(\tfrac{1}{2}\Sigma f^2 - \tfrac{1}{2}\Sigma\alpha^2\right). \tag{4}$$

Henceforward, having regard to the relation (3), δJ is independent of δF, and therefore of $\delta\alpha$. Let us now vary the other quantities.

The expression (1) for J gives

$$\delta J = \int dt \, d\tau (\Sigma f \, \delta f - \Sigma F \, \delta u).$$

But f, g, h must satisfy the first condition (2), so that

$$\Sigma \frac{d \, \delta f}{dx} = \delta\varrho, \tag{5}$$

and we may write

$$\delta J = \int dt \, d\tau \left[\Sigma f \, \delta f - \Sigma F \, \delta u - \psi\left(\Sigma \frac{d \, \delta f}{dx} - \delta\varrho\right)\right]. \tag{6}$$

From the calculus of variations, it is known that the calculation should be made as if ψ were an arbitrary function, δJ were represented by the expression (6), and the variations were not subject to the condition (5).

160 SPECIAL RELATIVITY

We also have

$$\delta u = \frac{d\,\delta f}{dt} + \delta(\varrho\xi),$$

and therefore, on integration by parts,

$$\delta J = \int dt\,d\tau \Sigma\,\delta f\left(f + \frac{dF}{dt} + \frac{d\psi}{dx}\right) + \int dt\,d\tau(\psi\,\delta\varrho - \Sigma F\,\delta(\varrho\xi)). \quad (7)$$

If now it be assumed that the electrons undergo no variation, then $\delta\varrho = \delta(\varrho\xi) = 0$, and the second integral vanishes. For δJ to be zero, we must have

$$f + \frac{dF}{dt} + \frac{d\psi}{dx} = 0. \quad (8)$$

In the general case, therefore,

$$\delta J = \int dt\,d\tau(\psi\,\delta\varrho - \Sigma F\,\delta(\varrho\xi)). \quad (9)$$

It remains to determine the forces acting upon the electrons. To do so, we must assume that a complementary force $-X\,d\tau$, $-Y\,d\tau$, $-Z\,d\tau$ is applied to each electron volume element, and write down the condition for this force to balance the forces of electromagnetic origin. Let U, V, W be the components of the displacement of the electron volume element $d\tau$, measured from any given initial position. Let δU, δV, δW be the variations of this displacement. The virtual work corresponding to the complementary force will be

$$-\int \Sigma X\,\delta U\,d\tau,$$

and the equilibrium condition just mentioned will therefore be

$$\delta J = -\int \Sigma X\,\delta U\,d\tau\,dt. \quad (10)$$

In order to transform δJ, we first seek the equation of continuity stating that the electron charge remains constant under the variation.

Let x_0, y_0, z_0 be the initial position of the electron. Its position at the time considered will be

$$x = x_0 + U, \quad y = y_0 + V, \quad z = z_0 + W.$$

We shall define also an auxiliary variable ε to generate the variation of each function: for any function A,

$$\delta A = \delta\varepsilon \frac{dA}{d\varepsilon}.$$

This is done because it will be convenient to be able to change between the notation of the calculus of variations and that of the ordinary differential calculus whenever desired.

The functions under consideration may be regarded in two ways: (a) as functions of the five variables x, y, z, t, ε, so that the position remains unaltered when only t and ε vary, in which case derivatives with be denoted by d as usual; (b) as functions of the five variables $x_0, y_0, z_0, t, \varepsilon$, so that a particular electron is followed when only t and ε vary, in which case derivatives will be denoted by the symbol ∂. Then we have

$$\xi = \frac{\partial U}{\partial t} = \frac{\partial U}{\partial t} + \xi\frac{dU}{dx} + \eta\frac{dU}{dy} + \zeta\frac{dU}{dz} = \frac{\partial x}{\partial t}. \qquad (11)$$

Now, let Δ denote the Jacobian of x, y, z with respect to x_0, y_0, z_0:

$$\Delta = \frac{\partial(x, y, z)}{\partial(x_0, y_0, z_0)}.$$

If t receives an increment ∂t while $\varepsilon, x_0, y_0, z_0$ remain constant, there will be consequent increments $\partial x, \partial y, \partial z$ of x, y, z, and $\partial\Delta$ of Δ, with

$$\partial x = \xi\,\partial t, \quad \partial y = \eta\,\partial t, \quad \partial z = \zeta\,\partial t,$$
$$\Delta + \partial\Delta = \frac{\partial(x+\partial x, y+\partial y, z+\partial z)}{\partial(x_0, y_0, z_0)}$$

whence
$$1+\frac{\partial \Delta}{\Delta} = \frac{\partial(x+\partial x, y+\partial y, z+\partial z)}{\partial(x, y, z)}$$
$$= \frac{\partial(x+\xi\,\partial t, y+\eta\,\partial t, z+\zeta\,\partial t)}{\partial(x, y, z)}.$$

From this we obtain
$$\frac{1}{\Delta}\frac{\partial \Delta}{\partial t} = \frac{d\xi}{dx}+\frac{d\eta}{dy}+\frac{d\zeta}{dz}. \tag{12}$$

Since the mass of an electron is constant,
$$\frac{\partial(\varrho\Delta)}{\partial t} = 0, \tag{13}$$
and therefore
$$\frac{\partial \varrho}{\partial t}+\Sigma\varrho\frac{d\xi}{dx} = 0, \quad \frac{\partial \varrho}{\partial t} = \frac{d\varrho}{dt}+\Sigma\xi\frac{d\varrho}{dx},$$
$$\frac{d\varrho}{dt}+\Sigma\frac{d(\varrho\xi)}{dx} = 0.$$

These are the various forms of the equation of continuity with respect to the variable t. Similar forms can be deduced with respect to the variable ε. Let
$$\delta U = \frac{\partial U}{\partial \varepsilon}\delta\varepsilon, \quad \delta V = \frac{\partial V}{\partial \varepsilon}\delta\varepsilon, \quad \delta W = \frac{\partial W}{\partial \varepsilon}\delta\varepsilon;$$
then
$$\delta U = \frac{dU}{d\varepsilon}\delta\varepsilon+\delta U\frac{dU}{dx}+\delta V\frac{dU}{dy}+\delta W\frac{dU}{dz}, \tag{11'}$$
$$\frac{1}{\Delta}\frac{\partial \Delta}{\partial \varepsilon} = \Sigma\frac{\partial U}{\partial \varepsilon}, \quad \frac{\partial(\varrho\Delta)}{\partial \varepsilon} = 0, \tag{12'}$$
$$\left.\begin{array}{c}\delta\varepsilon\dfrac{\partial \varrho}{\partial \varepsilon}+\Sigma\varrho\dfrac{d\delta U}{dx} = 0, \quad \dfrac{\partial \varrho}{\partial \varepsilon} = \dfrac{d\varrho}{d\varepsilon}+\Sigma\dfrac{\partial U}{\partial \varepsilon}\dfrac{d\varrho}{dx}, \\ \delta\varrho+\Sigma\dfrac{d\varrho\,\delta U}{dx} = 0.\end{array}\right\} \tag{13'}$$

It will be seen that there is a difference between the definition of $\delta U = (\partial U/\partial \varepsilon)\delta\varepsilon$ and that of $\delta\varrho = (d\varrho/d\varepsilon)\delta\varepsilon$, and that this definition of δU is the one which is appropriate to the formula (10).

The first term in equation (9) can be transformed by means of the last equation (13'):

$$\int dt\, d\tau \psi\, \delta\varrho = -\int dt\, d\tau \psi \Sigma \frac{d\varrho\, dU}{dx},$$

or, after integration by parts,

$$\int dt\, d\tau \psi\, \delta\varrho = \int dt\, d\tau\, \Sigma \varrho \frac{d\psi}{dx} \delta U. \tag{14}$$

Let us now seek to determine

$$\delta(\varrho\xi) = \frac{d(\varrho\xi)}{d\varepsilon} \delta\varepsilon.$$

We may notice that $\varrho\varDelta$ can depend only on x_0, y_0, z_0; for, if an electron volume element be considered whose initial position is a rectangular parallelepiped with edges dx_0, dy_0, dz_0, the charge on this element is

$$\varrho\varDelta\, dx_0\, dy_0\, dz_0.$$

Since the charge must remain constant,

$$\frac{\partial(\varrho\varDelta)}{\partial t} = \frac{\partial(\varrho\varDelta)}{\partial \varepsilon} = 0. \tag{15}$$

Hence we have

$$\frac{\partial^2(\varrho\, \varDelta U)}{\partial t\, \partial \varepsilon} = \frac{\partial}{\partial \varepsilon}\left(\varrho\varDelta \frac{\partial U}{\partial t}\right) = \frac{\partial}{\partial t}\left(\varrho\varDelta \frac{\partial U}{\partial \varepsilon}\right). \tag{16}$$

For any function A we have, by the equation of continuity,

$$\frac{1}{\varDelta} \frac{\partial(A\varDelta)}{\partial t} = \frac{dA}{dt} + \Sigma \frac{d(A\xi)}{dx}$$

and similarly

$$\frac{1}{\Delta} \frac{\partial(A\Delta)}{\partial \varepsilon} = \frac{dA}{d\varepsilon} + \Sigma \frac{d(A\,\partial U/\partial \varepsilon)}{dx}.$$

Hence

$$\frac{1}{\Delta} \frac{\partial}{\partial \varepsilon}\left(\varrho\Delta\,\frac{\partial U}{\partial t}\right) = \frac{d(\varrho\,\partial U/\partial t)}{d\varepsilon} + \frac{d(\varrho(\partial U/\partial t)(\partial U/\partial \varepsilon))}{dx}$$
$$+ \frac{d(\varrho(\partial U/\partial t)(\partial V/\partial \varepsilon))}{dy} + \frac{d(\varrho(\partial U/\partial t)(\partial W/\partial \varepsilon))}{dz}, \quad (17)$$

$$\frac{1}{\Delta} \frac{\partial}{\partial t}\left(\varrho\Delta\,\frac{\partial U}{\partial \varepsilon}\right) = \frac{d(\varrho\,\partial U/\partial \varepsilon)}{dt} + \frac{d(\varrho(\partial U/\partial t)(\partial U/\partial \varepsilon))}{dx}$$
$$+ \frac{d(\varrho(\partial V/\partial t)(\partial U/\partial \varepsilon))}{dy} + \frac{d(\varrho(\partial W/\partial t)(\partial U/\partial \varepsilon))}{dz}. \quad (17')$$

The right-hand sides of (17) and (17′) must be equal, and, since

$$\frac{\partial U}{\partial t} = \xi, \quad \frac{\partial U}{\partial \varepsilon}\,\delta\varepsilon = \delta U, \quad \frac{d(\varrho\xi)}{d\varepsilon}\,\delta\varepsilon = \delta(\varrho\xi),$$

we obtain

$$\delta(\varrho\xi) + \frac{d(\varrho\xi\,\delta U)}{dx} + \frac{d(\varrho\xi\,\delta V)}{dy} + \frac{d(\varrho\xi\,\delta W)}{dz}$$
$$= \frac{d(\varrho\,\delta U)}{dt} + \frac{d(\varrho\xi\,\delta U)}{dx} + \frac{d(\varrho\eta\,\delta U)}{dy} + \frac{d(\varrho\zeta\,\delta U)}{dz}. \quad (18)$$

Now transforming the second term in (9), we have

$$\int dt\,d\tau\,\Sigma F\,\delta(\varrho\xi)$$
$$= \int dt\,d\tau\bigg[\Sigma F\frac{d(\varrho\,\delta U)}{dt} + \Sigma F\frac{d(\varrho\eta\,\delta U)}{dy} + \Sigma F\frac{d(\varrho\zeta\,\delta U)}{dz}$$
$$-\Sigma F\frac{d(\varrho\xi\,\delta V)}{dy} - \Sigma F\frac{d(\varrho\xi\,\delta W)}{dz}\bigg].$$

Integration by parts on the right-hand side yields

$$\int dt\, d\tau \left[-\Sigma_\varrho\, \delta U \frac{dF}{dt} - \Sigma_\varrho\eta\, \delta U \frac{dF}{dy} - \Sigma_\varrho\zeta\, \delta U \frac{dF}{dz} \right.$$
$$\left. + \Sigma_\varrho\xi\, \delta V \frac{dF}{dy} + \Sigma_\varrho\xi\, \delta W \frac{dF}{dz} \right].$$

Next we note that

$$\Sigma_\varrho\xi\, \delta V \frac{dF}{dy} = \Sigma_\varrho\zeta\, \delta U \frac{dH}{dx},$$

$$\Sigma_\varrho\xi\, \delta W \frac{dF}{dz} = \Sigma_\varrho\eta\, \delta U \frac{dG}{dx}.$$

For, if the sums on either side are expanded, they become identities. Since also

$$\frac{dH}{dx} - \frac{dF}{dz} = -\beta, \quad \frac{dG}{dx} - \frac{dF}{dy} = \gamma,$$

the right-hand side becomes

$$\int dt\, d\tau \left[-\Sigma_\varrho\, \delta U \frac{dF}{dt} + \Sigma_\varrho\gamma\eta\, \delta U - \Sigma_\varrho\beta\zeta\, \delta U \right],$$

and thus finally

$$\delta J = \int dt\, d\tau\, \Sigma_\varrho\, \delta U \left(\frac{d\psi}{dx} + \frac{dF}{dt} + \beta\zeta - \gamma\eta \right)$$
$$= \int dt\, d\tau\, \Sigma_\varrho\, \delta U (-f + \beta\zeta - \gamma\eta).$$

Equating the coefficients of δU on either side of (10), we have

$$X = f - \beta\zeta + \gamma\eta.$$

This is equation (2) of Section 1.

§ 3. The Lorentz Transformation and the Principle of Least Action

Let us consider whether the principle of least action can explain the success of the Lorentz transformation. First of all, we must examine the result of applying this transformation to the integral

$$J = \int dt \, d\tau \left(\tfrac{1}{2} \Sigma f^2 - \tfrac{1}{2} \Sigma \alpha^2 \right)$$

(formula (4) of section 2).

We have firstly

$$dt' \, d\tau' = l^4 \, dt \, d\tau,$$

since x', y', z', t' are related to x, y, z, t by linear expressions whose determinant is l^4. Next,

$$\left. \begin{array}{l} l^4 \Sigma f'^2 = f^2 + k^2(g^2 + h^2) + k^2 \varepsilon^2 (\beta^2 + \gamma^2) + 2k^2 \varepsilon (g\gamma - h\beta), \\ l^4 \Sigma \alpha'^2 = \alpha^2 + k^2 (\beta^2 + \gamma^2) + k^2 \varepsilon^2 (g^2 + h^2) + 2k^2 \varepsilon (g\gamma - h\beta) \end{array} \right\} \quad (1)$$

(formulae (9) of section 1), whence

$$l^4 (\Sigma f'^2 - \Sigma \alpha'^2) = \Sigma f^2 - \Sigma \alpha^2.$$

Thus, if we put

$$J' = \int dt' \, d\tau' \left(\tfrac{1}{2} \Sigma f'^2 - \tfrac{1}{2} \Sigma \alpha'^2 \right),$$

the result is

$$J' = J.$$

However, for this equation to be valid, the limits of integration must be the same. Hitherto we have assumed that t ranged from t_0 to t_1, and x, y, z from $-\infty$ to $+\infty$. The limits of integration would then be altered by the Lorentz transformation; but there is no bar to assuming that $t_0 = -\infty$, $t_1 = +\infty$, and the limits for J and for J' are then the same.

We have thus to compare the two following equations, which are analogues of equation (10) in section 2:

$$\left.\begin{aligned}\delta J &= -\int \Sigma X\ \delta U\ d\tau\ dt, \\ \delta J' &= -\int \Sigma X'\ \delta U'\ d\tau'\ dt'.\end{aligned}\right\} \quad (2)$$

To do so, we must first compare $\delta U'$ with δU.

Let us consider an electron having initial co-ordinates x_0, y_0, z_0. Its co-ordinates at the instant t will be

$$x = x_0+U, \quad y = y_0+V, \quad z = z_0+W.$$

If the corresponding electron after the Lorentz transformation is considered, its co-ordinates will be

$$x' = kl(x+\varepsilon t), \quad y' = ly, \quad z' = lz,$$

where

$$x' = x_0+U', \quad y' = y_0+V', \quad z' = z_0+W';$$

but these values will be reached at the instant

$$t' = kl(t+\varepsilon x).$$

If the variables are subjected to variations δU, δV, δW, while at the same time t receives an increment δt, then the total increments of the co-ordinates x, y, z will be

$$\delta x = \delta U+\xi\ \delta t, \quad \delta y = \delta V+\eta\ \delta t, \quad \delta z = \delta W+\zeta\ \delta t.$$

Similarly,

$$\delta x' = \delta U'+\xi'\ \delta t',\ \delta y' = \delta V'+\eta'\ \delta t',\ \delta z' = \delta W'+\zeta'\ \delta t',$$

and, by the Lorentz transformation,

$$\delta x' = kl(\delta x+\varepsilon\ \delta t), \quad \delta y' = l\ \delta y, \quad \delta z' = l\ \delta z,$$
$$\delta t' = kl(\delta t+\varepsilon\ \delta x);$$

hence, assuming $\delta t = 0$, we find

$$\delta x' = \delta U' + \xi' \, \delta t' = kl \, \delta U,$$
$$\delta y' = \delta V' + \eta' \, \delta t' = l \, \delta V,$$
$$\delta t' = kl\varepsilon \, \delta U.$$

Since

$$\xi' = \frac{\xi + \varepsilon}{1 + \xi\varepsilon}, \quad \eta' = \frac{\eta}{k(1 + \xi\varepsilon)},$$

we have, on replacing $\delta t'$ by its value,

$$kl(1 + \xi\varepsilon) \, \delta U = \delta U'(1 + \xi\varepsilon) + (\xi + \varepsilon)kl\varepsilon \, \delta U,$$
$$l(1 + \xi\varepsilon) \, \delta V = \delta V'(1 + \xi\varepsilon) + \eta l\varepsilon \, \delta U.$$

Using the definition of k, we obtain from these equations

$$\delta U = \frac{k}{l} \delta U' + \frac{k\varepsilon}{l} \xi \, \delta U',$$

$$\delta V = \frac{1}{l} \delta V' + \frac{k\varepsilon}{l} \eta \, \delta U',$$

and similarly

$$\delta W = \frac{1}{l} \delta W' + \frac{k\varepsilon}{l} \zeta \, \delta U';$$

hence

$$\Sigma X \, \delta U = \frac{1}{l}(kX \, \delta U' + Y \, \delta V' + Z \, \delta W') + \frac{k\varepsilon}{l} \delta U' \Sigma X \xi. \quad (3)$$

Now, according to the equations (2), we must have

$$\int \Sigma X' \, \delta U' \, dt' \, d\tau' = \int \Sigma X \, \delta U \, dt \, d\tau = \frac{1}{l^4} \int \Sigma X \, \delta U \, dt' \, d\tau'.$$

Replacing $\Sigma X \, \delta U$ by its value (3) and equating coefficients, we find

$$X' = \frac{k}{l^5} X + \frac{k\varepsilon}{l^5} \Sigma X \xi, \quad Y' = \frac{1}{l^5} Y, \quad Z' = \frac{1}{l^5} Z.$$

These are the equations (11) of Section 1. Thus the principle of least action leads to the same results as does the analysis given in Section 1.

Returning to formulae (1), we see that $\Sigma f^2 - \Sigma \alpha^2$ is unaltered by the Lorentz transformation, apart from a constant factor. The same is not true of the expression $\Sigma f^2 + \Sigma \alpha^2$ which appears in the energy. If we consider only the case where ε is so small that its square may be neglected, so that $k = 1$, and if we also assume that $l = 1$, then

$$\Sigma f'^2 = \Sigma f^2 + 2\varepsilon(g\gamma - h\beta),$$
$$\Sigma \alpha'^2 = \Sigma \alpha^2 + 2\varepsilon(g\gamma - h\beta),$$

and, by addition,

$$\Sigma f'^2 + \Sigma \alpha'^2 = \Sigma f^2 + \Sigma \alpha^2 + 4\varepsilon(g\gamma - h\beta).$$

§ 4. The Lorentz Group

It is noteworthy that the Lorentz transformations form a group. For, if we put

$$x' = kl(x + \varepsilon t), \quad y' = ly, \quad z' = lz, \quad t' = kl(t + \varepsilon x),$$

and

$$x'' = k'l'(x' + \varepsilon' t'), \quad y'' = l'y', \quad z'' = l'z', \quad t'' = k'l'(t' + \varepsilon' x'),$$

with

$$k^{-2} = 1 - \varepsilon^2, \quad k'^{-2} = 1 - \varepsilon'^2,$$

we find that

$$x'' = k''l''(x + \varepsilon'' t), \quad y'' = l''y,$$
$$z'' = l''z, \quad t'' = k''l''(t + \varepsilon'' x),$$

with

$$\varepsilon'' = \frac{\varepsilon + \varepsilon'}{1 + \varepsilon \varepsilon'}, \quad l'' = ll', \quad k'' = kk'(1 + \varepsilon \varepsilon') = \frac{1}{\sqrt{(1 - \varepsilon''^2)}}.$$

Taking $l = 1$ and assuming ε infinitesimal, with

$$x' = x + \delta x, \quad y' = y + \delta y, \quad z' = z + \delta z, \quad t' = t + \delta t,$$

we have
$$\delta x = \varepsilon t, \quad \delta y = \delta z = 0, \quad \delta t = \varepsilon x.$$

This is the infinitesimal generating transformation of the group, which I shall denote by T_1, and which in Lie's notation may be written

$$t\frac{d\phi}{dx} + x\frac{d\phi}{dt} = T_1.$$

If we take $\varepsilon = 0$ and $l = 1 + \delta l$, on the other hand, we obtain

$$\delta x = x\,\delta l, \quad \delta y = y\,\delta l, \quad \delta z = z\,\delta l, \quad \delta t = t\,\delta l,$$

which yields another infinitesimal transformation T_0 of the group (assuming that l and ε are regarded as independent variables); in Lie's notation,

$$T_0 = x\frac{d\phi}{dx} + y\frac{d\phi}{dy} + z\frac{d\phi}{dz} + t\frac{d\phi}{dt}.$$

It is also possible to assign to the y-axis or to the z-axis the particular significance which has been given to the x-axis, thus obtaining two further infinitesimal transformations

$$T_2 = t\frac{d\phi}{dy} + y\frac{d\phi}{dt},$$

$$T_3 = t\frac{d\phi}{dz} + z\frac{d\phi}{dt},$$

which likewise would leave Lorentz's equations unchanged.

The combinations defined by Lie, such as

$$[T_1, T_2] = x\frac{d\phi}{dy} - y\frac{d\phi}{dx},$$

can also be constructed; but it is easily seen that this transformation is equivalent to a rotation of the co-ordinate axes through a very small angle about the z-axis. It is therefore not surprising that

POINCARÉ: THE DYNAMICS OF THE ELECTRON 171

this does not affect the form of Lorentz's equations, which are obviously independent of the axes chosen.

We are thus led to consider a continuous group, to be called the *Lorentz group*, possessing the following infinitesimal transformations:

(1) the transformation T_0, which commutes with every other;
(2) the three transformations T_1, T_2, T_3;
(3) the three rotations $[T_1, T_2], [T_2, T_3], [T_3, T_1]$.

Any transformation belonging to this group can be resolved into a transformation having the form

$$x' = lx, \quad y' = ly, \quad z' = lz, \quad t' = lt$$

and a linear transformation which leaves unaltered the quadratic form

$$x^2 + y^2 + z^2 - t^2.$$

The group can also be generated in another way. Any transformation of the group may be regarded as comprising a transformation having the form

$$x' = kl(x + \varepsilon t), \quad y' = ly, \quad z' = lz, \quad t' = kl(t + \varepsilon x), \quad (1)$$

preceded and followed by an appropriate rotation.

For our purposes, however, we have to consider only certain of the transformations in this group. We must regard l as being a function of ε, the function being chosen so that this partial group, which will be denoted by P, is itself a group.

Let the system be rotated through 180° about the y-axis; then the resulting transformation must also belong to P. This operation is equivalent to changing the signs of x, x', z and z'; hence we have

$$x' = kl(x - \varepsilon t), \quad y' = ly, \quad z' = lz, \quad t' = kl(t - \varepsilon x). \quad (2)$$

Thus l is unchanged when ε is replaced by $-\varepsilon$.

Next, if P is a group, the substitution inverse to (1), which is

$$x' = \frac{k}{l}(x - \varepsilon t), \quad y' = \frac{y}{l}, \quad z' = \frac{z}{l}, \quad t' = \frac{k}{l}(t - \varepsilon x), \quad (3)$$

must likewise belong to P; it must therefore be identical with (2), so that

$$l = 1/l.$$

Consequently, we must have $l = 1$.

> [Note: there follow here four sections of a technical character, dealing with details of the electron theory not now of importance. Then Poincaré continues:]

§ 9. Hypotheses Concerning Gravitation

Thus Lorentz' theory would entirely account for the impossibility of demonstrating absolute motion, provided that all forces were of electromagnetic origin.

But there exist forces, such as gravitation, which cannot be regarded as being of electromagnetic origin. It may happen that two systems of bodies create equivalent electromagnetic fields, in the sense of exerting the same action upon electrified bodies and currents, while at the same time these two systems do not exert the same gravitational action upon Newtonian masses. The gravitational field is therefore not identical with the electromagnetic field. Lorentz was thus compelled to augment his hypothesis by assuming that *forces, of whatever origin, and in particular gravitation, are affected by translation* (or, if one prefers, by the Lorentz transformation) *in the same way as the electromagnetic forces.*

We must now examine this hypothesis in detail. If the Newtonian force is to behave in such a way under the Lorentz transformation, we can no longer suppose that this force depends only on the relative position of the attracting and the attracted body at the instant concerned; it must depend also on the velocities of

POINCARÉ: THE DYNAMICS OF THE ELECTRON 173

the two bodies. Moreover, we may reasonably assume that the force acting upon the attracted body, at an instant t, depends on the position and velocity of the body at that instant; but it will also depend on the position and velocity of the *attracting* body, not at the instant t but at some *previous* instant, as if gravitation required a certain time for its propagation.

Let us consider therefore the position of the attracted body at the instant t_0, and let its co-ordinates at that instant be x_0, y_0, z_0, and the components of its velocity be ξ, η, ζ; and let us consider the attracting body at the corresponding instant t_0+t, its co-ordinates at that instant being x_0+x, y_0+y, z_0+z, and its velocity components ξ_1, η_1, ζ_1.

First of all, we must have a relationship

$$\phi(t, x, y, z, \xi, \eta, \zeta, \xi_1, \eta_1, \zeta_1) = 0 \tag{1}$$

to determine the time t. This relationship expresses the law of propagation of gravitational action; I shall by no means impose the condition that propagation occurs with the same velocity in every direction.

Next, let X_1, Y_1, Z_1 be the three components of the action exerted upon the attracted body at the instant t. We have to express X_1, Y_1, Z_1 as functions of

$$t, x, y, z, \xi, \eta, \zeta, \xi_1, \eta_1, \zeta_1. \tag{2}$$

The conditions to be satisfied are as follows.

1. The relationship (1) must not be affected by the transformations of the Lorentz group.

2. The components X_1, Y_1, Z_1 must behave, under the Lorentz transformations, in the same manner as the electromagnetic forces denoted by the same letters, that is, as shown by equations (11′) of Section 1.

3. When both bodies are at rest, the usual law of attraction must apply.

In the latter case, however, it should be noted that the relationship (1) plays no part, since the time t is of no significance if both bodies are at rest.

The problem thus stated is clearly indeterminate. We shall therefore seek to satisfy as many further conditions as possible.

4. Astronomical observations do not appear to reveal any perceptible deviation from Newton's law, and we shall therefore choose the solution which differs least from this law when the velocities of the two bodies are small.

5. We shall attempt to ensure that t is always negative; for, whereas it is reasonable that the effect of gravitation should require a certain time for its propagation, we should find it more difficult to understand how this effect could depend on a position of the attracting body which the latter has *not yet reached*.

There is one case where the problem is no longer indeterminate, namely if the two bodies are at *relative* rest, i.e. if

$$\xi = \xi_1, \quad \eta = \eta_1, \quad \zeta = \zeta_1;$$

we shall therefore first investigate this case, assuming that these velocities are constant, and therefore that the two bodies are executing a common uniform motion of translation in a straight line.

We may assume that the x-axis has been taken to be parallel to this motion of translation, so that $\eta = \zeta = 0$, and we shall take $\varepsilon = -\xi$.

If, under these conditions, we apply the Lorentz transformation, the two bodies will be at rest after the transformation, with

$$\xi' = \eta' = \zeta' = 0$$

The components X_1', Y_1', Z_1' must then be in accordance with Newton's law and we have, apart from a constant factor,

$$\left. \begin{array}{c} X_1' = -\dfrac{x'}{r'^3}, \quad Y_1' = -\dfrac{y'}{r'^3}, \quad Z_1' = -\dfrac{z'}{r'^3}, \\ r'^2 = x'^2 + y'^2 + z'^2. \end{array} \right\} \quad (3)$$

But, from Section 1,

$$x' = k(x+\varepsilon t), \quad y' = y, \quad z' = z, \quad t' = k(t+\varepsilon x),$$

$$\frac{\varrho'}{\varrho} = k(1+\xi\varepsilon) = k(1-\varepsilon^2) = \frac{1}{k}, \quad \Sigma X_1\xi = -X_1\varepsilon,$$

$$X_1' = k\frac{\varrho}{\varrho'}(X_1+\varepsilon\Sigma X_1\xi) = k^2 X_1(1-\varepsilon^2) = X_1,$$

$$Y_1' = \frac{\varrho}{\varrho'}Y_1 = kY_1,$$

$$Z_1' = kZ_1.$$

Moreover,

$$x+\varepsilon t = x-\xi t, \quad r'^2 = k^2(x-\xi t)^2 + y^2 + z^2,$$

and

$$X_1 = \frac{-k(x-\xi t)}{r'^3}, \quad Y_1 = \frac{-y}{kr'^3}, \quad Z_1 = \frac{-z}{kr'^3}, \quad (4)$$

which may also be written

$$X_1 = \frac{dV}{dx}, \quad Y_1 = \frac{dV}{dy}, \quad Z_1 = \frac{dV}{dz}; \quad V = \frac{1}{kr'}. \quad (4')$$

It seems at first sight that the indeterminacy remains, since no hypotheses have been made concerning the value of t, that is, concerning the velocity of propagation. Moreover, x is a function of t. But it is easily seen that the quantities $x-\xi t$, y and z which appear in the formulae do not depend on t.

Thus, if the two bodies have a common translatory motion, the force acting upon the attracted body is normal to an ellipsoid having the attracting body at its centre.

In order to proceed further, it is necessary to ascertain the *invariants of the Lorentz group*.

It is known that the substitutions forming this group (if $l = 1$) are linear and such that the quadratic form

$$x^2+y^2+z^2-t^2$$

is invariant. Putting

$$\xi = \frac{\delta x}{\delta t}, \qquad \eta = \frac{\delta y}{\delta t}, \qquad \zeta = \frac{\delta z}{\delta t};$$

$$\xi_1 = \frac{\delta_1 x}{\delta_1 t}, \qquad \eta_1 = \frac{\delta_1 y}{\delta_1 t}, \qquad \zeta_1 = \frac{\delta_1 z}{\delta_1 t},$$

we see that the Lorentz transformation causes δx, δy, δz, δt and $\delta_1 x$, $\delta_1 y$, $\delta_1 z$, $\delta_1 t$ to undergo the same linear substitutions as x, y, z, t.

If

$$\begin{array}{cccc} x & y & z & t\sqrt{-1} \\ \delta x & \delta y & \delta z & \delta t\sqrt{-1} \\ \delta_1 x & \delta_1 y & \delta_1 z & \delta_1 t\sqrt{-1} \end{array}$$

are regarded as the co-ordinates of three points P, P', P'' in four-dimensional space, we see that the Lorentz transformation is simply a rotation of this space about a fixed origin. The only distinct invariants are therefore the six distances of the points P, P', P'' from one another and from the origin, or alternatively the two expressions

$$x^2 + y^2 + z^2 - t^2, \quad x\,\delta x + y\,\delta y + z\,\delta z - t\,\delta t$$

and the four expressions of the same form obtained by permuting the three points P, P', P'' in any manner.

What we are seeking, however, is invariant functions of the ten variables (2); we must therefore find, among combinations of the six invariants, those which depend only on these ten variables, i.e. those which are homogeneous and of degree zero with respect to δx, δy, δz, δt and with respect to $\delta_1 x$, $\delta_1 y$, $\delta_1 z$, $\delta_1 t$. This leaves four distinct invariants, namely

$$\Sigma x^2 - t^2, \quad \frac{t - \Sigma x\xi}{\sqrt{(1 - \Sigma \xi^2)}}, \quad \frac{t - \Sigma x\xi_1}{\sqrt{(1 - \Sigma \xi_1^2)}}, \quad \frac{1 - \Sigma \xi \xi_1}{\sqrt{[(1 - \Sigma \xi^2)(1 - \Sigma \xi_1^2)]}}. \quad (5)$$

Let us now consider how the components of the force are transformed. We return to equations (11) of Section 1, which refer

not to the force X_1, Y_1, Z_1 discussed here but to the force X, Y, Z per unit volume. Putting
$$T = \Sigma X\xi,$$
we see that these equations (11) may be written (with $l = 1$)

$$\left. \begin{array}{ll} X' = k(X+\varepsilon T), & T' = k(T+\varepsilon X), \\ Y' = Y, & Z' = Z; \end{array} \right\} \quad (6)$$

thus, X, Y, Z, T are transformed in the same manner as x, y, z, t. The invariants of the group will therefore be

$$\Sigma X^2 - T^2, \quad \Sigma Xx - Tt, \quad \Sigma X\,\delta x - T\,\delta t, \quad \Sigma X\,\delta_1 x - T\,\delta_1 t.$$

The quantities in which we are interested are not X, Y, Z, but X_1, Y_1, Z_1, with
$$T_1 = \Sigma X_1 \xi.$$
Evidently
$$\frac{X_1}{X} = \frac{Y_1}{Y} = \frac{Z_1}{Z} = \frac{T_1}{T} = \frac{1}{\varrho}.$$

Thus the Lorentz transformation will act upon X_1, Y_1, Z_1, T_1 in the same way as upon X, Y, Z, T, except that these expressions will in addition be multiplied by

$$\frac{\varrho}{\varrho'} = \frac{1}{k(1+\xi\varepsilon)} = \frac{\delta t}{\delta t'}.$$

Likewise, the transformation will act upon $\xi, \eta, \zeta, 1$ in the same way as upon $\delta x, \delta y, \delta z, \delta t$, except that these expressions will in addition be multiplied by the *same* factor,

$$\frac{\delta t}{\delta t'} = \frac{1}{k(1+\xi\varepsilon)}.$$

Let us now regard $X, Y, Z, T\sqrt{-1}$ as being the co-ordinates of a fourth point Q; the invariants will then be functions of the distances between the five points

$$O, P, P', P'', Q;$$

and these functions must be homogeneous of degree zero, firstly with respect to

$$X, Y, Z, T, \quad \delta x, \quad \delta y, \quad \delta z, \quad \delta t$$

(which variables can subsequently be replaced by X_1, Y_1, Z_1, T_1, $\xi, \eta, \zeta, 1$), and secondly with respect to

$$\delta_1 x, \quad \delta_1 y, \quad \delta_1 z, \quad 1$$

(which variables can subsequently be replaced by $\xi_1, \eta_1, \zeta_1, 1$).

In this way we find, in addition to the four invariants (5), four further and distinct invariants, namely

$$\frac{\Sigma X_1^2 - T_1^2}{1 - \Sigma \xi^2}, \quad \frac{\Sigma X_1 x - T_1 t}{\sqrt{(1 - \Sigma \xi^2)}}, \quad \frac{\Sigma X_1 \xi_1 - T_1}{\sqrt{[(1 - \Sigma \xi^2)(1 - \Sigma \xi_1^2)]}}, \quad \frac{\Sigma X_1 \xi - T_1}{1 - \Sigma \xi^2}. \tag{7}$$

The last of these is always zero, according to the definition of T_1.

Which are the conditions that must now be satisfied?

1. The left-hand side of equation (1), which defines the velocity of propagation, must be a function of the four invariants (5).

It is obvious that a large number of hypotheses could be constructed. We shall consider only two of these.

(A) It may be that
$$\Sigma x^2 - t^2 = r^2 - t^2 = 0,$$

whence $t = \pm r$; and, since t must be negative, $t = -r$. This means that the velocity of propagation is equal to that of light. At first sight, it seems that this hypothesis should be rejected immediately; for Laplace has shown that the propagation is either instantaneous or much more rapid than that of light. But Laplace was discussing the hypothesis of a finite velocity of propagation alone, whereas here it is compounded with many others, and there may happen to be some more or less complete mutual compensation between them, a situation of which many examples have already appeared in the applications of the Lorentz transformation.

(B) It may be that
$$\frac{t-\Sigma x\xi_1}{\sqrt{(1-\Sigma\xi_1^2)}} = 0, \quad t = \Sigma x\xi_1.$$

The velocity of propagation is then much more rapid than that of light, but in certain cases t might be negative, which, as we have said, seems hardly acceptable. *We shall therefore abide by hypothesis* (A).

2. The four invariants (7) must be functions of the invariants (5).

3. When both bodies are at absolute rest, X_1, Y_1, Z_1 must have the values given by Newton's law; when the bodies are at relative rest, the values must be those given by equations (4).

In the case of absolute rest, the first two invariants (7) must reduce to
$$\Sigma X_1^2, \quad \Sigma X_1 x,$$
or, by Newton's law, to
$$1/r^4, \quad -1/r.$$

According to hypothesis (A), the second and third of the invariants (5) become
$$\frac{-r-\Sigma x\xi}{\sqrt{(1-\Sigma\xi^2)}}, \quad \frac{-r-\Sigma x\xi_1}{\sqrt{(1-\Sigma\xi_1^2)}},$$
that is, for absolute rest,
$$-r, \quad -r.$$

We may therefore assume, *for example*, that the first two invariants (5) reduce to
$$\frac{(1-\Sigma\xi_1^2)^2}{(r+\Sigma x\xi_1)^4} - \frac{\sqrt{(1-\Sigma\xi_1^2)}}{r+\Sigma x\xi_1};$$
but other combinations are possible.

It is necessary to choose some combination, and a third equation is also needed in order to determine X_1, Y_1, Z_1. In making the choice, we shall attempt to remain as close as possible to Newton's law. Let us then examine the result when the squares of the velocities ξ, η, etc., are neglected (and $t = -r$). The four invariants (5) then become

$$0, \quad -r - \Sigma x\xi, \quad -r - \Sigma x\xi_1, \quad 1,$$

and the four invariants (7) become

$$\Sigma X_1^2, \quad \Sigma X_1(x + \xi r), \quad \Sigma X_1(\xi_1 - \xi), \quad 0.$$

In order to compare this with Newton's law, however, a further transformation is necessary. In these equations, $x_0 + x$, $y_0 + y$, $z_0 + z$ represent the co-ordinates of the attracting body at the instant $t_0 + t$, and $r = \sqrt{\Sigma x^2}$; in Newton's law, we have to consider the co-ordinates $x_0 + x_1$, $y_0 + y_1$, $z_0 + z_1$ of the attracting body at the instant t_0, and the distance $r_1 = \sqrt{\Sigma x_1^2}$.

We may neglect the square of the time t occupied by the propagation, and therefore regard the motion as uniform; then

$$x = x_1 + \xi_1 t, \quad y = y_1 + \eta_1 t, \quad z = z_1 + \zeta_1 t,$$
$$r(r - r_1) = \Sigma x \xi_1 t;$$

or, since $t = -r$,

$$x = x_1 - \xi_1 r, \quad y = y_1 - \eta_1 r, \quad z = z_1 - \zeta_1 r, \quad r = r_1 - \Sigma x \xi_1,$$

and the four invariants (5) become

$$0, \quad -r_1 + \Sigma x(\xi_1 - \xi), \quad -r_1, \quad 1$$

and the four invariants (7)

$$\Sigma X_1^2, \quad \Sigma X_1[x_1 + (\xi - \xi_1) r_1], \quad \Sigma X_1(\xi_1 - \xi), \quad 0.$$

In the second of these expressions I have written r_1 in place of r, since r is multiplied by $\xi - \xi_1$ and the square of ξ is neglected.

Newton's law gives, for these four invariants (7),

$$\frac{1}{r_1^4}, \quad -\frac{1}{r_1} - \frac{\Sigma x_1(\xi - \xi_1)}{r_1^2}, \quad \frac{\Sigma x_1(\xi - \xi_1)}{r_1^3}, \quad 0.$$

If therefore we denote the second and third invariants (5) by A and B, and the first three invariants (7) by M, N and P, Newton's law will be obeyed, to within terms of the order of the squares of the velocities, by putting

$$M = \frac{1}{B^4}, \quad N = \frac{+A}{B}, \quad P = \frac{A-B}{B^3}. \tag{8}$$

This solution is not unique: if the fourth invariant (5) is denoted by C, then $C-1$ is of the order of ξ^2, as is $(A-B)^2$. We may therefore add to the right-hand side of each of the equations (8) a term consisting of $C-1$ multiplied by any function of A, B and C, and a term consisting of $(A-B)^2$ also multiplied by any function of A, B and C.

The solution (8) appears the simplest at first sight, but it cannot be accepted. Since M, N and P are functions of X_1, Y_1, Z_1 and $T_1 = \Sigma X_1 \xi$, these equations yield values of X_1, Y_1 and Z_1; but the resulting values may in some cases be imaginary.

In order to avoid this difficulty, we proceed differently, putting

$$k_0 = \frac{1}{\sqrt{(1-\Sigma\xi^2)}}, \quad k_1 = \frac{1}{\sqrt{(1-\Sigma\xi_1^2)}},$$

by analogy with

$$k = \frac{1}{\sqrt{(1-\varepsilon^2)}},$$

as in the Lorentz substitution.

Then, with the condition $-r = t$, the invariants (5) become

$$0, \quad A = -k_0(r+\Sigma x\xi), \quad B = -k_1(r+\Sigma x\xi_1),$$
$$C = k_0 k_1 (1-\Sigma\xi\xi_1).$$

Moreover, the following systems of quantities:

x,	y,	z,	$-r = t$
$k_0 X_1$,	$k_0 Y_1$,	$k_0 Z_1$,	$k_0 T_1$
$k_0 \xi$,	$k_0 \eta$,	$k_0 \zeta$,	k_0
$k_1 \xi_1$,	$k_1 \eta_1$,	$k_1 \zeta_1$,	k_1

are seen to undergo the *same* linear substitutions when the transformations of the Lorentz group are applied to them. We therefore put

$$\left.\begin{aligned}X_1 &= x\frac{\alpha}{k_0} + \xi\beta + \xi_1 \frac{k_1}{k_0}\gamma,\\ Y_1 &= y\frac{\alpha}{k_0} + \eta\beta + \eta_1 \frac{k_1}{k_0}\gamma,\\ Z_1 &= z\frac{\alpha}{k_0} + \zeta\beta + \zeta_1 \frac{k_1}{k_0}\gamma,\\ T_1 &= -r\frac{\alpha}{k_0} + \beta + \frac{k_1}{k_0}\gamma.\end{aligned}\right\} \quad (9)$$

It is evident that, if α, β, γ are invariants, X_1, Y_1, Z_1, T_1 will satisfy the fundamental condition, i.e. will undergo an appropriate linear substitution when the Lorentz transformations are applied to them.

If the equations (9) are compatible, we must have

$$\Sigma X_1 \xi - T_1 = 0.$$

When X_1, Y_1, Z_1, T_1 are replaced by their values (9), the result is, after multiplication by k_0^2,

$$-A\alpha - \beta - C\gamma = 0. \quad (10)$$

The desired conclusion is that the values of X_1, Y_1, Z_1 should remain in accordance with Newton's law when the squares of the velocities ξ, etc., and the products of the accelerations and the distances are neglected in comparison with the square of the velocity of light.

We can take

$$\beta = 0, \quad \gamma = -A\alpha/C.$$

To the approximation used,

$$k_0 = k_1 = 1, \quad C = 1, \quad A = -r_1 + \Sigma x(\xi_1 - \xi),$$
$$B = -r_1, \quad x = x_1 + \xi_1 t = x_1 - \xi_1 r.$$

Then the first equation (9) becomes

$$X_1 = \alpha(x - A\xi_1).$$

But, if ξ^2 is neglected, $A\xi_1$ may be replaced by $-r_1\xi_1$, or by $-r\xi_1$, whence

$$X_1 = \alpha(x + \xi_1 r) = \alpha x_1.$$

Newton's law would give

$$X_1 = -x_1/r_1^3.$$

We must therefore take as the invariant α one which reduces to $-1/r_1^3$ within the approximation adopted, that is, $1/B^3$. The equations (9) then become

$$\left.\begin{aligned}
X_1 &= \frac{x}{k_0 B^3} - \xi_1 \frac{k_1}{k_0} \frac{A}{B^3 C}, \\
Y_1 &= \frac{y}{k_0 B^3} - \eta_1 \frac{k_1}{k_0} \frac{A}{B^3 C}, \\
Z_1 &= \frac{z}{k_0 B^3} - \zeta_1 \frac{k_1}{k_0} \frac{A}{B^3 C}, \\
T_1 &= -\frac{r}{k_0 B^3} - \frac{k_1}{k_0} \frac{A}{B^3 C}.
\end{aligned}\right\} \quad (11)$$

It is seen, first of all, that the corrected attraction consists of two components, one parallel to the vector joining the positions of the two bodies, and the other parallel to the velocity of the attracting body.

When we speak of the position or the velocity of the attracting body, we mean its position or velocity at the instant when the gravitational wave leaves it; but the position or the velocity of the attracted body means its position or velocity at the instant when the gravitational wave reaches it, this wave being assumed to be propagated with the velocity of light.

I believe that it would be premature to attempt to continue the discussion of these formulae, and I shall therefore confine myself to making a few comments.

1. The solutions (11) are not unique; for the common factor $1/B^3$ may be replaced by

$$\frac{1}{B^3} + (C-1)f_1(A, B, C) + (A-B)^2 f_2(A, B, C),$$

where f_1 and f_2 are any functions of A, B and C. Moreover, β need not be taken as zero; any additional terms may be added to α, β and γ which satisfy the condition (10) and are of the second order n ξ for α, and of the first order in ξ for β and γ.

2. The first equation (11) may be written

$$X_1 = \frac{k_1}{B^3 C}[x(1 - \Sigma\xi\xi_1) + \xi_1(r + \Sigma x\xi)], \qquad (11')$$

and the quantity in the brackets may in turn be written

$$(x + r\xi_1) + \eta(\xi_1 y - x\eta_1) + \zeta(\xi_1 z - x\zeta_1), \qquad (12)$$

so that the total force is divisible into three components corresponding to the three parentheses in equation (12). The first component is somewhat similar to the mechanical force due to the electric field, the other two to the mechanical force due to the magnetic field. By virtue of comment 1, I may replace $1/B^3$ in equations (11) by C/B^3, so that X_1, Y_1, Z_1 are linear functions of the velocity ξ, η, ζ of the attracted body, C having been eliminated from the denominator of (11'). This completes the analogy.

Putting then

$$\left.\begin{array}{lll} k_1(x+r\xi_1) = \lambda, & k_1(y+r\eta_1) = \mu, & k_1(z+r\zeta_1) = \nu, \\ k_1(\eta_1 z - \zeta_1 y) = \lambda', & k_1(\zeta_1 x - \xi_1 z) = \mu', & k_1(\xi_1 y - x\eta_1) = \nu', \end{array}\right\} \qquad (13)$$

with C eliminated from the denominator of (11′) we obtain

$$\left.\begin{aligned} X_1 &= \frac{\lambda}{B^3} + \frac{\eta \nu' - \zeta \mu'}{B^3}, \\ Y_1 &= \frac{\mu}{B^3} + \frac{\zeta \lambda' - \xi \nu'}{B^3}, \\ Z_1 &= \frac{\nu}{B^3} + \frac{\xi \mu' - \eta \lambda'}{B^3}, \end{aligned}\right\} \qquad (14)$$

and also

$$B^2 = \Sigma \lambda^2 - \Sigma \lambda'^2. \qquad (15)$$

Thus λ, μ, ν or λ/B^3, μ/B^3, ν/B^3 is a kind of electric field, while λ', μ', ν' or λ'/B^3, μ'/B^3, ν'/B^3 is a kind of magnetic field.

3. The relativity postulate would compel us to use either the solution (11) or the solution (14) or any one of the solutions obtained therefrom by using comment 1. But the prime question is whether these are compatible with astronomical observations. The deviation from Newton's law is of the order of ξ^2, that is, 10,000 times less than if it had been of the order of ξ, as it would have been with the velocity of propagation equal to that of light and the other conditions unchanged. We may therefore hope that the deviation will not be very great; but only a more extended investigation will furnish the answer to this question.

Paris

July 1905

NOTES ON EXTRACT 5

IN THIS absolutely astonishing paper of Einstein's we find no mention of the simultaneous work of Lorentz and Poincaré, and it is clear that Einstein has worked entirely independently. He goes right to the root of matters in the conception of the invariance of Maxwell's theory, by starting with the electrodynamic interaction between a magnet and a current. Contrasting the effect when the current moves and the magnet is at rest with that when the magnet moves and the current is at rest, then going on to consider the question of the transmission of light, he relates at once two apparently different problems: (i) the difficulty of a one-way determination of the speed of light, (ii) the question of a rest system for Maxwell's equations. All the elementary consequences of the Lorentz transformation are worked out as soon as it has been found, and Einstein then goes on to show that the Maxwell equations are indeed invariant under these transformations, which have been found in a way which is more or less independent of Maxwell's equations. He realises that mechanics will need some modification in the light of his results, and so he reformulates mechanics in the later part of the paper.

EXTRACT 5

On the Electrodynamics of Moving Bodies

By A. EINSTEIN

Translated from "Zur Electrodynamik bewegter Körper," *Annalen der Physik*, **17**, 891 (1905).

IT IS known that Maxwell's electrodynamics—as usually understood at the present time—when applied to moving bodies, leads to asymmetries which do not appear to be inherent in the phenomena. Take, for example, the reciprocal electrodynamic action of a magnet and a conductor. The observable phenomenon here depends only on the relative motion of the conductor and the magnet, whereas the customary view draws a sharp distinction between the two cases in which either the one or the other of these bodies is in motion. For if the magnet is in motion and the conductor at rest, there arises in the neighbourhood of the magnet an electric field with a certain definite energy, producing a current at the places where parts of the conductor are situated. But if the magnet is stationary and the conductor in motion, no electric field arises in the neighbourhood of the magnet. In the conductor, however, we find an electromotive force, to which in itself there is no corresponding energy, but which gives rise—assuming equality of relative motion in the two cases discussed—to electric currents of the same path and intensity as those produced by the electric forces in the former case.

Examples of this sort, together with the unsuccessful attempts to discover any motion of the earth relatively to the "light medium," suggest that the phenomena of electrodynamics as well as of mechanics possess no properties corresponding to the idea of absolute rest. They suggest rather that, as has already been shown to the first order of small quantities, the same laws of electrodynamics and optics will be valid for all frames of reference for which the equations of mechanics hold good. We will raise this conjcture (the purport of which will hereafter be called the "Principle of Relativity") to the status of a postulate, and also introduce another postulate, which is only apparently irreconcilable with the former, namely, that light is always propagated in empty space with a definite velocity c which is independent of the state of motion of the emitting body. These two postulates suffice for the attainment of a simple and consistent theory of the electrodynamics of moving bodies based on Maxwell's theory for stationary bodies. The introduction of a "luminiferous ether" will prove to be superfluous inasmuch as the view here to be developed will not require an "absolutely stationary space" provided with special properties, nor assign a velocity-vector to a point of the empty space in which electromagnetic processes take place.

The theory to be developed is based—like all electrodynamics—on the kinematics of the rigid body, since the assertions of any such theory have to do with the relationships between rigid bodies (systems of co-ordinates), clocks, and electromagnetic processes. Insufficient consideration of this circumstance lies at the root of the difficulties which the electrodynamics of moving bodies at present encounters.

I. Kinematical Part

§ 1. *Definition of Simultaneity*

Let us take a system of co-ordinates in which the equations of Newtonian mechanics hold good. In order to render our presentation more precise and to distinguish this system of co-ordinates verbally from others which will be introduced hereafter, we call it the "stationary system."

If a material point is at rest relatively to this system of co-ordinates, its position can be defined relatively thereto by the employment of rigid standards of measurement and the methods of Euclidean geometry, and can be expressed in Cartesian co-ordinates.

If we wish to describe the *motion* of a material point, we give the values of its co-ordinates as functions of the time. Now we must bear carefully in mind that a mathematical description of this kind has no physical meaning unless we are quite clear as to what we understand by "time." We have to take into account that all our judgments in which time plays a part are always judgments of *simultaneous events*. If, for instance, I say, "That train arrives here at 7 o'clock," I mean something like this: "The pointing of the small hand of my watch to 7 and the arrival of the train are simultaneous events."*

It might appear possible to overcome all the difficulties attending the definition of "time" by substituting "the position of the small hand of my watch" for "time." And in fact such a definition is satisfactory when we are concerned with defining a time exclusively for the place where the watch is located; but it is no longer satisfactory when we have to connect in time series of events occurring at different places, or—what comes to the same thing—to evaluate the times of events occurring at places remote from the watch.

We might, of course, content ourselves with time values determined by an observer stationed together with the watch at the origin of the co-ordinates, and co-ordinating the corresponding positions of the hands with light signals, given out by every event to be timed, and reaching him through empty space. But this co-ordination has the disadvantage that it is not independent of the standpoint of the observer with the watch or clock, as we know

* We shall not here discuss the inexactitude which lurks in the concept of simultaneity of two events at approximately the same place, which can only be removed by an abstraction.

from experience. We arrive at a much more practical determination along the following line of thought.

If at the point A of space there is a clock, an observer at A can determine the time values of events in the immediate proximity of A by finding the positions of the hands which are simultaneous with these events. If there is at the point B of space another clock in all respects resembling the one at A, it is possible for an observer at B to determine the time values of events in the immediate neighbourhood of B. But it is not possible without further assumption to compare, in respect of time, an event at A with an event at B. We have so far defined only an "A time" and a "B time." We have not defined a common "time" for A and B. The latter time can now be defined in establishing *by definition* that the "time" required by light to travel from A to B equals the "time" it requires to travel from B to A. Let a ray of light start at the "A time" t_A from A towards B, let it at the "B time" t_B be reflected at B in the direction of A, and arrive again at A at the "A time" t'_A.

In accordance with definition the two clocks synchronize if

$$t_B - t_A = t'_A - t_B.$$

We assume that this definition of synchronism is free from contradictions, and possible for any number of points; and that the following relations are universally valid:

1. If the clock at B synchronizes with the clock at A, the clock at A synchronizes with the clock at B.
2. If the clock at A synchronizes with the clock at B and also with the clock at C, the clocks at B and C also synchronize with each other.

Thus with the help of certain imaginary physical experiments we have settled what is to be understood by synchronous stationary clocks located at different places, and have evidently obtained a definition of "simultaneous," or "synchronous," and of "time." The "time" of an event is that which is given simultaneously with the event by a stationary clock located at the place of the event, this clock being synchronous, and indeed synchronous for all time determinations, with a specified stationary clock.

In agreement with experience we further assume the quantity

$$\frac{2AB}{t'_A - t_A} = c,$$

to be a universal constant—the velocity of light in empty space.

An essential point is that we have defined time by means of stationary clocks in the stationary system, and the time now defined being appropriate to the stationary system we call it "the time of the stationary system."

§ 2. *On the Relativity of Lengths and Times*

The following reflexions are based on the principle of relativity and on the principle of the constancy of the velocity of light. These two principles we define as follows:

1. The laws by which the states of physical systems undergo change are not affected, whether these changes of state be referred to the one or the other of two systems of co-ordinates in uniform translatory motion relative to each other.

2. Any ray of light moves in the "stationary" system of co-ordinates with the determined velocity c, whether the ray be emitted by a stationary or by a moving body. Here

$$\text{velocity} = \frac{\text{light path}}{\text{time interval}}$$

where time interval is to be taken in the sense of the definition in § 1.

Let there be given a stationary rigid rod; and let its length be l as measured by a measuring-rod which is also stationary. We now imagine the axis of the rod lying along the axis of x of the stationary system of co-ordinates, and that a uniform motion of parallel translation with velocity v along the axis of x in the direction of increasing x is then imparted to the rod. We now inquire as to the length of the moving rod, and imagine its length to be ascertained by the following two operations:

(*a*) The observer moves together with the given measuring-rod and the rod to be measured, and measures the length of the rod directly by superposing the measuring-rod, in just the same way as if all three were at rest.

(*b*) By means of stationary clocks set up in the stationary system and synchronizing in accordance with § 1, the observer ascertains at what points of the stationary system the two ends of the rod to be measured are located at a definite time. The distance between these two points, measured by the measuring-rod already employed, which in this case is at rest, is also a length which may be designated "the length of the rod."

In accordance with the principle of relativity the length to be discovered by the operation (*a*)—we will call it "the length of the rod in the moving system"—must be equal to the length l of the stationary rod.

The length to be discovered by the operation (*b*) we will call "the length of the (moving) rod in the stationary system." This we shall determine on the basis of our two principles, and we shall find that it differs from l.

Current kinematics tacitly assumes that the lengths determined by these two operations are precisely equal, or in other words, that a moving rigid body at the epoch t may in geometrical respects be perfectly represented by *the same* body *at rest* in a definite position.

We imagine further that at the two ends A and B of the rod, clocks are placed which synchronize with the clocks of the stationary system, that is to say that their indications correspond at any instant to the "time of the stationary system" at the places where they happen to be. These clocks are therefore "synchronous in the stationary system."

We imagine further that with each clock there is a moving observer, and that these observers apply to both clocks the criterion established in § 1 for the synchronization of two clocks. Let a ray of light depart from A at the time* t_A, let it be reflected at B at the time t_B, and reach A again at the time t'_A. Taking into consideration the principle of the constancy of the velocity of light we find that

$$t_B - t_A = \frac{r_{AB}}{c-v} \quad \text{and} \quad t'_A - t_B = \frac{r_{AB}}{c+v}$$

where r_{AB} denotes the length of the moving rod—measured in the stationary system. Observers moving with the moving rod would thus find that the two clocks were not synchronous, while observers in the stationary system would declare the clocks to be synchronous.

So we see that we cannot attach any *absolute* signification to the concept of simultaneity, but that two events which, viewed from a system of co-ordinates, are simultaneous, can no longer be looked upon as simultaneous events when envisaged from a system which is in motion relatively to that system.

§ 3. *Theory of the Transformation of Co-ordinates and Times from a Stationary System to another System in Uniform Motion of Translation Relatively to the Former*

Let us in "stationary" space take two systems of co-ordinates, i.e. two systems, each of three rigid material lines, perpendicular to one another, and issuing from a point. Let the axes of X of the two systems coincide, and their axes of Y and Z respectively be parallel. Let each system be provided with a rigid measuring-rod and a number of clocks, and let the two measuring-rods, and likewise all the clocks of the two systems, be in all respects alike.

---

* "Time" here denotes "time of the stationary system" and also "position of hands of the moving clock situated at the place under discussion."

Now to the origin of one of the two systems (k) let a constant velocity v be imparted in the direction of increasing x of the other stationary system (K), and let this velocity be communicated to the axes of the co-ordinates, the relevant measuring-rod, and the clocks. To any time t of the stationary system K there then will correspond a definite position of the axes of the moving system, and from reasons of symmetry we are entitled to assume that the motion of k may be such that the axes of the moving system are at the time t (this "t" always denotes a time of the stationary system) parallel to the axes of the stationary system.

We now imagine space to be measured from the stationary system K by means of the stationary measuring-rod, and also from the moving system k by means of the measuring-rod moving with it; and that we thus obtain the co-ordinates x, y, z, and ξ, η, ζ respectively. Further, let the time t of the stationary system be determined by means of the clocks at rest in the stationary system, for all points thereof at which there are clocks, by means of light signals in the manner indicated in § 1; similarly let the time τ of the moving system be determined for all points of the moving system at which there are clocks at rest relatively to that system by applying the method, given in § 1, of light signals between the points at which the latter clocks are located.

To any system of values x, y, z, t, which completely defines the place and time of an event in the stationary system, there belongs a system of values ξ, η, ζ, τ, determining that event relatively to the system k, and our task is now to find the system of equations connecting these quantities.

In the first place it is clear that the equations must be *linear* on account of the properties of homogeneity which we attribute to space and time.

If we place $x' = x - vt$, it is clear that a point at rest in the system k must have a system of values x', y, z, independent of time. We first determine τ as a function of x', y, z, and t. To do this we have to express in equations that τ is nothing else than the summary of the data of clocks at rest in system k, which have been synchronized according to the rule given in § 1.

From the origin of system k let a ray be emitted at the time τ_0 along the X-axis to x', and at the time τ_1 be reflected thence to the origin of the co-ordinates, arriving there at the time τ_2; we then must have $\frac{1}{2}(\tau_0+\tau_2) = \tau_1$, or, by inserting the arguments of the function τ and applying the principle of the constancy of the velocity of light in the stationary system:

$$\frac{1}{2}\left[\tau(0,0,0,t)+\tau\left(0,0,0,t+\frac{x'}{c-v}+\frac{x'}{c+v}\right)\right] = \tau\left(x',0,0,t+\frac{x'}{c-v}\right).$$

Hence, if x' be chosen infinitesimally small,

$$\frac{1}{2}\left(\frac{1}{c-v}+\frac{1}{c+v}\right)\frac{\partial \tau}{\partial t} = \frac{\partial \tau}{\partial x'}+\frac{1}{c-v}\frac{\partial \tau}{\partial t},$$

or

$$\frac{\partial \tau}{\partial x'}+\frac{v}{c^2-v^2}\frac{\partial \tau}{\partial t} = 0.$$

It is to be noted that instead of the origin of the co-ordinates we might have chosen any other point for the point of origin of the ray, and the equation just obtained is therefore valid for all values of x', y, z.

An analogous consideration—applied to the axes of H and Z—it being borne in mind that light is always propagated along these axes, when viewed from the stationary system, with the velocity $\sqrt{(c^2-v^2)}$, gives us

$$\frac{\partial \tau}{\partial y} = 0, \quad \frac{\partial \tau}{\partial z} = 0.$$

Since τ is a *linear* function, it follows from these equations that

$$\tau = a\left(t-\frac{v}{c^2-v^2}x'\right)$$

where a is a function $\phi(v)$ at present unknown, and where for brevity it is assumed that at the origin of k, $t = 0$, when $\tau = 0$.

With the help of this result we easily determine the quantities ξ, η, ζ by expressing in equations that light (as required by the principle of the constancy of the velocity of light, in combination with the principle of relativity) is also propagated with velocity c when measured in the moving system. For a ray of light emitted at the time $\tau = 0$ in the direction of the increasing ξ

$$\xi = c\tau \quad \text{or} \quad \xi = ac\left(t - \frac{v}{c^2 - v^2} x'\right).$$

But the ray moves relatively to the origin of k, when measured in the stationary system, with the velocity $c-v$, so that

$$\frac{x'}{c-v} = t.$$

If we insert this value of t in the equation for ξ, we obtain

$$\xi = a \frac{c^2}{c^2 - v^2} x'.$$

In an analogous manner we find, by considering rays moving along the two other axes, that

$$\eta = c\tau = ac\left(t - \frac{v}{c^2 - v^2} x'\right)$$

where

$$\frac{y}{\sqrt{(c^2 - v^2)}} = t, \quad x' = 0.$$

Thus

$$\eta = a \frac{c}{\sqrt{(c^2 - v^2)}} y \quad \text{and} \quad \zeta = a \frac{c}{\sqrt{(c^2 - v^2)}} z.$$

Substituting for x' its value, we obtain

$$\tau = \phi(v)\beta(t - vx/c^2),$$
$$\xi = \phi(v)\beta(x - vt),$$
$$\eta = \phi(v)y,$$
$$\zeta = \phi(v)z,$$

where

$$\beta = \frac{1}{\sqrt{(1-v^2/c^2)}},$$

and ϕ is an as yet unknown function of v. If no assumption whatever be made as to the initial position of the moving system and as to the zero point of τ, an additive constant is to be placed on the right side of each of these equations.

We now have to prove that any ray of light, measured in the moving system, is propagated with the velocity c, if, as we have assumed, this is the case in the stationary system; for we have not as yet furnished the proof that the principle of the constancy of the velocity of light is compatible with the principle of relativity.

At the time $t = \tau = 0$, when the origin of the co-ordinates is common to the two systems, let a spherical wave be emitted therefrom, and be propagated with the velocity c in system K. If (x, y, z) be a point just attained by this wave, then

$$x^2 + y^2 + z^2 = c^2 t^2.$$

Transforming this equation with the aid of our equations of transformation we obtain after a simple calculation

$$\xi^2 + \eta^2 + \zeta^2 = c^2 \tau^2.$$

The wave under consideration is therefore no less a spherical wave with velocity of propagation c when viewed in the moving system. This shows that our two fundamental principles are compatible.

In the equations of transformation which have been developed there enters an unknown function ϕ of v, which we will now determine.

For this purpose we introduce a third system of co-ordinates K', which relatively to the system k is in a state of parallel translatory motion parallel to the axis of Ξ, such that the origin of co-ordinates of system k moves with velocity $-v$ on the axis of Ξ. At the time $t = 0$ let all three origins coincide, and when $t = x = y = z = 0$ let the time t' of the system K' be zero. We call the co-ordinates, measured in the system K', x', y', z', and by a twofold application of our equations of transformation we obtain

$$\begin{aligned} t' &= \phi(-v)\beta(-v)(\tau+v\xi/c^2) &= \phi(v)\phi(-v)t, \\ x' &= \phi(-v)\beta(-v)(\xi+v\tau) &= \phi(v)\phi(-v)x, \\ y' &= \phi(-v)\eta &= \phi(v)\phi(-v)y, \\ z' &= \phi(-v)\zeta &= \phi(v)\phi(-v)z. \end{aligned}$$

Since the relations between x', y', z' and x, y, z do not contain the time t, the systems K and K' are at rest with respect to one another, and it is clear that the transformation from K to K' must be the identical transformation. Thus

$$\phi(v)\phi(-v) = 1.$$

We now inquire into the signification of $\phi(v)$. We give our attention to that part of the axis of Y of system k which lies between $\xi = 0$, $\eta = 0$, $\zeta = 0$ and $\xi = 0$, $\eta = l$, $\zeta = 0$. This part of the axis of H is a rod moving perpendicularly to its axis with velocity v relatively to system K. Its ends possess in K the co-ordinates

$$x_1 = vt, \quad y_1 = \frac{l}{\phi(v)}, \quad z_1 = 0$$

and

$$x_2 = vt, \quad y_2 = 0, \quad z_2 = 0.$$

The length of the rod measured in K is therefore $l/\phi(v)$; and this gives us the meaning of the function $\phi(v)$. From reasons of symmetry it is now evident that the length of a given rod moving perpendicularly to its axis, measured in the stationary system, must depend only on the velocity and not on the direction and the sense of the motion. The length of the moving rod measured in the stationary system does not change, therefore, if v and $-v$ are interchanged. Hence follows that $l/\phi(v) = l/\phi(-v)$, or

$$\phi(v) = \phi(-v).$$

It follows from this relation and the one previously found that $\phi(v) = 1$, so that the transformation equations which have been found become

$$\tau = \beta(t - vx/c^2),$$
$$\xi = \beta(x - vt),$$
$$\eta = y,$$
$$\zeta = z,$$

where

$$\beta = 1/\sqrt{(1 - v^2/c^2)}.$$

§ 4. Physical Meaning of the Equations Obtained in Respect to Moving Rigid Bodies and Moving Clocks

We envisage a rigid sphere* of radius R, at rest relatively to the moving system k, and with its centre at the origin of co-ordinates of k. The equation of the surface of this sphere moving relatively to the system K with velocity v is

$$\xi^2 + \eta^2 + \zeta^2 = R^2.$$

The equation of this surface expressed in x, y, z at the time $t = 0$ is

$$\frac{x^2}{(\sqrt{(1 - v^2/c^2)})^2} + y^2 + z^2 = R^2.$$

A rigid body which, measured in a state of rest, has the form of a sphere, therefore has in a state of motion—viewed from the stationary system—the form of an ellipsoid of revolution with the axes

$$R\sqrt{(1 - v^2/c^2)}, R, R.$$

Thus, whereas the Y and Z dimensions of the sphere (and therefore of every rigid body of no matter what form) do not appear modified by the motion, the X dimension appears shortened in the

---

* That is, a body possessing spherical form when examined at rest.

ratio $1 : \sqrt{(1-v^2/c^2)}$, i.e. the greater the value of v, the greater the shortening. For $v = c$ all moving objects—viewed from the "stationary" system—shrivel up into plane figures. For velocities greater than that of light our deliberations become meaningless; we shall, however, find in what follows, that the velocity of light in our theory plays the part, physically, of an infinitely great velocity.

It is clear that the same results hold good of bodies at rest in the "stationary" system, viewed from a system in uniform motion.

Further, we imagine one of the clocks which are qualified to mark the time t when at rest relatively to the stationary system, and the time τ when at rest relatively to the moving system, to be located at the origin of the co-ordinates of k, and so adjusted that it marks the time τ. What is the rate of this clock, when viewed from the stationary system?

Between the quantities x, t, and τ, which refer to the position of the clock, we have, evidently, $x = vt$ and

$$\tau = \frac{1}{\sqrt{(1-v^2/c^2)}} (t - vx/c^2).$$

Therefore,

$$\tau = t\sqrt{(1-v^2/c^2)} = t - \left(1 - \sqrt{(1-v^2/c^2)}\right)t$$

whence it follows that the time marked by the clock (viewed in the stationary system) is slow by $1 - \sqrt{(1-v^2/c^2)}$ seconds per second, or—neglecting magnitudes of fourth and higher order—by $\frac{1}{2}v^2/c^2$.

From this there ensues the following peculiar consequence. If at the points A and B of K there are stationary clocks which, viewed in the stationary system, are synchronous; and if the clock at A is moved with the velocity v along the line AB to B, then on its arrival at B the two clocks no longer synchronize, but the clock moved from A to B lags behind the other which has remained at B by $\frac{1}{2}tv^2/c^2$ (up to magnitudes of fourth and higher order), t being the time occupied in the journey from A to B.

It is at once apparent that this result still holds good if the clock moves from A to B in any polygonal line, and also when the points A and B coincide.

If we assume that the result proved for a polygonal line is also valid for a continuously curved line, we arrive at this result: If one of two synchronous clocks at A is moved in a closed curve with constant velocity until it returns to A, the journey lasting t seconds, then by the clock which has remained at rest the travelled clock on its arrival at A will be $\frac{1}{2}tv^2/c^2$ second slow. Thence we conclude that a balance-clock* at the equator must go more slowly, by a very small amount, than a precisely similar clock situated at one of the poles under otherwise identical conditions.

§ 5. *The Composition of Velocities*

In the system k moving along the axis of X of the system K with velocity v, let a point move in accordance with the equations

$$\xi = w_\xi \tau, \quad \eta = w_\eta \tau, \quad \zeta = 0,$$

where w_ξ and w_η denote constants.

Required: the motion of the point relatively to the system K. If with the help of the equations of transformation developed in § 3 we introduce the quantities x, y, z, t into the equations of motion of the point, we obtain

$$x = \frac{w_\xi + v}{1 + vw_\xi/c^2} t,$$

$$y = \frac{\sqrt{(1-v^2/c^2)}}{1 + vw_\xi/c^2} w_\eta t,$$

$$z = 0.$$

Thus the law of the parallelogram of velocities is valid accord-

* Not a pendulum-clock, which is physically a system to which the Earth belongs. This case had to be excluded.

ing to our theory only to a first approximation. We set

$$V^2 = \left(\frac{dx}{dt}\right)^2 + \left(\frac{dy}{dt}\right)^2,$$
$$w^2 = w_\xi^2 + w_\eta^2,$$
$$a = \tan^{-1} w_y/w_x,$$

a is then to be looked upon as the angle between the velocities v and w. After a simple calculation we obtain

$$V = \frac{\sqrt{[(v^2 + w^2 + 2vw \cos a) - (vw \sin a/c^2)^2]}}{1 + vw \cos a/c^2}.$$

It is worthy of remark that v and w enter into the expression for the resultant velocity in a symmetrical manner. If w also has the direction of the axis of X (axis of Ξ), we get

$$V = \frac{v+w}{1+vw/c^2}.$$

It follows from this equation that from a composition of two velocities which are less than c, there always results a velocity less than c. For if we set $v = c - \varkappa$, $w = c - \lambda$, $\varkappa$ and λ being positive and less than c, then

$$V = c \frac{2c - \varkappa - \lambda}{2c - \varkappa - \lambda + \varkappa\lambda/c} < c.$$

It follows, further, that the velocity of light c cannot be altered by composition with a velocity less than that of light. For this case we obtain

$$V = \frac{c+w}{1+w/c} = c.$$

We might also have obtained the formula for V, for the case when v and w have the same direction, by compounding two transformations in accordance with § 3. If in addition to the systems K and k figuring in § 3 we introduce still another system of co-ordinates k' moving parallel to k, its origin moving on the axis of X

with the velocity w, we obtain equations between the quantities x, y, z, t and the corresponding quantities of k', which differ from the equations found in § 3 only in that the place of "v" is taken by the quantity

$$\frac{v+w}{1+vw/c^2};$$

from which we see that such parallel transformations form a group, as they must do.

We have now deduced the requisite laws of the theory of kinematics corresponding to our two principles, and we proceed to show their application to electrodynamics.

II. Electrodynamical Part

§ 6. Transformation of the Maxwell–Hertz Equations for Empty Space. On the Nature of the Electromotive Forces Occurring in a Magnetic Field During Motion

Let the Maxwell–Hertz equations for empty space hold good for the stationary system K, so that we have

$$\frac{1}{c}\frac{\partial X}{\partial t} = \frac{\partial N}{\partial y} - \frac{\partial M}{\partial z}, \quad \frac{1}{c}\frac{\partial L}{\partial t} = \frac{\partial Y}{\partial z} - \frac{\partial Z}{\partial y},$$

$$\frac{1}{c}\frac{\partial Y}{\partial t} = \frac{\partial L}{\partial z} - \frac{\partial N}{\partial x}, \quad \frac{1}{c}\frac{\partial M}{\partial t} = \frac{\partial Z}{\partial x} - \frac{\partial X}{\partial z},$$

$$\frac{1}{c}\frac{\partial Z}{\partial t} = \frac{\partial M}{\partial x} - \frac{\partial L}{\partial y}, \quad \frac{1}{c}\frac{\partial N}{\partial t} = \frac{\partial X}{\partial y} - \frac{\partial Y}{\partial x},$$

where (X, Y, Z) denotes the vector of the electric force, and (L, M, N) that of the magnetic force.

If we apply to these equations the transformation developed in § 3, by referring the electromagnetic processes to the system of coordinates there introduced, moving with the velocity v, we obtain

204 SPECIAL RELATIVITY

the equations

$$\frac{1}{c}\frac{\partial X}{\partial \tau} = \frac{\partial}{\partial \eta}\left\{\beta\left(N - \frac{v}{c}Y\right)\right\} - \frac{\partial}{\partial \zeta}\left\{\beta\left(M + \frac{v}{c}Z\right)\right\},$$

$$\frac{1}{c}\frac{\partial}{\partial \tau}\left\{\beta\left(Y - \frac{v}{c}N\right)\right\} = \frac{\partial L}{\partial \zeta} \qquad\qquad - \frac{\partial}{\partial \xi}\left\{\beta\left(N - \frac{v}{c}Y\right)\right\}.$$

$$\frac{1}{c}\frac{\partial}{\partial \tau}\left\{\beta\left(Z + \frac{v}{c}M\right)\right\} = \frac{\partial}{\partial \xi}\left\{\beta\left(M + \frac{v}{c}Z\right)\right\} - \frac{\partial L}{\partial \eta},$$

$$\frac{1}{c}\frac{\partial L}{\partial \tau} = \frac{\partial}{\partial \zeta}\left\{\beta\left(Y - \frac{v}{c}N\right\} - \frac{\partial}{\partial \eta}\left\{\beta\left(Z + \frac{v}{c}M\right)\right\},$$

$$\frac{1}{c}\frac{\partial}{\partial \tau}\left\{\beta\left(M + \frac{v}{c}Z\right)\right\} = \frac{\partial}{\partial \xi}\left\{\beta\left(Z + \frac{v}{c}M\right)\right\} - \frac{\partial X}{\partial \zeta},$$

$$\frac{1}{c}\frac{\partial}{\partial \tau}\left\{\beta\left(N - \frac{v}{c}Y\right)\right\} = \frac{\partial X}{\partial \eta} \qquad\qquad - \frac{\partial}{\partial \xi}\left\{\beta\left(Y - \frac{v}{c}N\right)\right\},$$

where

$$\beta = 1/\sqrt{(1 - v^2/c^2)}.$$

Now the principle of relativity requires that if the Maxwell–Hertz equations for empty space hold good in system K, they also hold good in system k; that is to say that the vectors of the electric and the magnetic force—(X', Y', Z') and (L', M', N')—of the moving system k, which are defined by their ponderomotive effects on electric or magnetic charges respectively, satisfy the following equations:

$$\frac{1}{c}\frac{\partial X'}{\partial \tau} = \frac{\partial N'}{\partial \eta} - \frac{\partial M'}{\partial \zeta}, \qquad \frac{1}{c}\frac{\partial L'}{\partial \tau} = \frac{\partial Y'}{\partial \zeta} - \frac{\partial Z'}{\partial \eta},$$

$$\frac{1}{c}\frac{\partial Y'}{\partial \tau} = \frac{\partial L'}{\partial \zeta} - \frac{\partial N'}{\partial \xi}, \qquad \frac{1}{c}\frac{\partial M'}{\partial \tau} = \frac{\partial Z'}{\partial \xi} - \frac{\partial X'}{\partial \zeta},$$

$$\frac{1}{c}\frac{\partial Z'}{\partial \tau} = \frac{\partial M'}{\partial \xi} - \frac{\partial L'}{\partial \eta}, \qquad \frac{1}{c}\frac{\partial N'}{\partial \tau} = \frac{\partial X'}{\partial \eta} - \frac{\partial Y'}{\partial \xi}.$$

Evidently the two systems of equations found for system k must express exactly the same thing, since both systems of equations are equivalent to the Maxwell–Hertz equations for system K. Since, further, the equations of the two systems agree, with the exception of the symbols for the vectors, it follows that the functions occurring in the systems of equations at corresponding places must agree, with the exception of a factor $\psi(v)$, which is common for all functions of the one system of equations, and is independent of ξ, η, ζ and τ but may depend upon v. Thus we have the relations

$$X' = \psi(v)X, \qquad L' = \psi(v)L,$$
$$Y' = \psi(v)\beta\left(Y - \frac{v}{c}N\right), \quad M' = \psi(v)\beta\left(M + \frac{v}{c}Z\right),$$
$$Z' = \psi(v)\beta\left(Z + \frac{v}{c}M\right), \quad N' = \psi(v)\beta\left(N - \frac{v}{c}Y\right).$$

If we now form the reciprocal of this system of equations, firstly by solving the equations just obtained, and secondly by applying the equations to the inverse transformation (from k to K), which is characterized by the velocity $-v$, it follows, when we consider that the two systems of equations thus obtained must be identical, that $\psi(v)\psi(-v) = 1$. Further, from reasons of symmetry* $\psi(v) = \psi(-v)$, and therefore

$$\psi(v) = 1,$$

and our equations assume the form

$$X' = X, \qquad L' = L,$$
$$Y' = \beta\left(Y - \frac{v}{c}N\right), \quad M' = \beta\left(M + \frac{v}{c}Z\right),$$
$$Z' = \beta\left(Z + \frac{v}{c}M\right), \quad N' = \beta\left(N - \frac{v}{c}Y\right).$$

As to the interpretation of these equations we make the follow-

* If, for example, $X = Y = Z = L = M = 0$, and $N \neq 0$, then from reasons of symmetry it is clear that when v changes sign without changing its numerical value, Y' must also change sign without changing its numerical value.

ing remarks: Let a point charge of electricity have the magnitude "one" when measured in the stationary system K, i.e. let it when at rest in the stationary system exert a force of one dyne upon an equal quantity of electricity at a distance of one cm. By the principle of relativity this electric charge is also of the magnitude "one" when measured in the moving system. If this quantity of electricity is at rest relatively to the stationary system, then by definition the vector (X, Y, Z) is equal to the force acting upon it. If the quantity of electricity is at rest relatively to the moving sysem (at least at the relevant instant), then the force acting upon it, measured in the moving system, is equal to the vector (X', Y', Z'). Consequently the first three equations above allow themselves to be clothed in words in the two following ways:

1. If a unit electric point charge is in motion in an electromagnetic field, there acts upon it, in addition to the electric force, an "electromotive force" which, if we neglect the terms multiplied by the second and higher powers of v/c, is equal to the vector-product of the velocity of the charge and the magnetic force, divided by the velocity of light. (Old manner of expression.)

2. If a unit electric point charge is in motion in an electromagnetic field, the force acting upon it is equal to the electric force which is present at the locality of the charge, and which we ascertain by transformation of the field to a system of co-ordinates at rest relatively to the electrical charge. (New manner of expression.)

An analogous situation holds with "magnetomotive forces." We see that electromotive force plays in the developed theory merely the part of an auxiliary concept, which owes its introduction to the circumstance that electric and magnetic forces do not exist independently of the state of motion of the system of co-ordinates.

Furthermore it is clear that the asymmetry mentioned in the introduction as arising when we consider the currents produced by the relative motion of a magnet and a conductor, now disappears. Moreover, questions as to the "seat" of electrodynamic electromotive forces (unipolar machines) now have no point.

§ 7. *Theory of Doppler's Principle and of Aberration*

In the system K, very far from the origin of co-ordinates, let there be a source of electrodynamic waves, which in a part of space containing the origin of co-ordinates may be represented to a sufficient degree of approximation by the equations

$$X = X_0 \sin \Phi, \quad L = L_0 \sin \Phi,$$
$$Y = Y_0 \sin \Phi, \quad M = M_0 \sin \Phi,$$
$$Z = Z_0 \sin \Phi, \quad N = N_0 \sin \Phi,$$

where

$$\Phi = \omega \left\{ t - \frac{1}{c}(lx + my + nz) \right\}.$$

Here (X_0, Y_0, Z_0) and (L_0, M_0, N_0) are the vectors defining the amplitude of the wave-train, and l, m, n the direction-cosines of the wave-normals. We wish to know the constitution of these waves, when they are examined by an observer at rest in the moving system k.

Applying the equations of transformation found in § 6 for electric and magnetic forces, and those found in § 3 for the co-ordinates and the time, we obtain directly

$$X' = X_0 \sin \Phi', \qquad L' = L_0 \sin \Phi',$$
$$Y' = \beta(Y_0 - vN_0/c) \sin \Phi', \quad M' = \beta(M_0 + vZ_0/c) \sin \Phi',$$
$$Z' = \beta(Z_0 + vM_0/c) \sin \Phi', \quad N' = \beta(N_0 - vY_0/c) \sin \Phi'.$$
$$\Phi' = \omega' \left\{ \tau - \frac{1}{c}(l'\xi + m'\eta + n'\zeta) \right\}$$

where

$$\omega' = \omega\beta(1 - lv/c),$$
$$l' = \frac{l - v/c}{1 - lv/c},$$
$$m' = \frac{m}{\beta(1 - lv/c)},$$
$$n' = \frac{n}{\beta(1 - lv/c)}.$$

From the equation for ω' it follows that if an observer is moving with velocity v relatively to an infinitely distant source of light of frequency ν, in such a way that the connecting line "source—observer" makes the angle ϕ with the velocity of the observer referred to a system of co-ordinates which is at rest relatively to the source of light, the frequency ν' of the light perceived by the observer is given by the equation

$$\nu' = \nu \frac{1-\cos\phi \cdot v/c}{\sqrt{(1-v^2/c^2)}}.$$

This is Doppler's principle for any velocities whatever. When $\phi = 0$ the equation assumes the perspicuous form

$$\nu' = \nu \sqrt{\frac{1-v/c}{1+v/c}}.$$

We see that, in contrast with the customary view, when $v = -c$, $\nu' = \infty$.

If we call the angle between the wave-normal (direction of the ray) in the moving system and the connecting line "source—observer" ϕ', the equation for l' assumes the form

$$\cos\phi' = \frac{\cos\phi - v/c}{1 - \cos\phi \cdot v/c}.$$

This equation expresses the law of aberration in its most general form. If $\phi = \frac{1}{2}\pi$, the equation becomes simply

$$\cos\phi' = -v/c.$$

We still have to find the amplitude of the waves, as it appears in the moving system. If we call the amplitude of the electric or magnetic force A or A' respectively, accordingly as it is measured in the stationary system or in the moving system, we obtain

$$A'^2 = A^2 \frac{(1-\cos\phi \cdot v/c)^2}{1-v^2/c^2}$$

which equation, if $\phi = 0$, simplifies into

$$A'^2 = A^2 \frac{1-v/c}{1+v/c}.$$

It follows from these results that to an observer approaching a source of light with the velocity c, this source of light must appear of infinite intensity.

§ 8. Transformation of the Energy of Light Rays. Theory of the Pressure of Radiation Exerted on Perfect Reflectors

Since $A^2/8\pi$ equals the energy of light per unit of volume, we have to regard $A'^2/8\pi$, by the principle of relativity, as the energy of light in the moving system. Thus A'^2/A^2 would be the ratio of the "measured in motion" to the "measured at rest" energy of a given light complex, if the volume of a light complex were the same, whether measured in K or in k. But this is not the case. If l, m, n are the direction-cosines of the wave-normal of the light in the stationary system, no energy passes through the surface elements of a spherical surface moving with the velocity of light:

$$(x-lct)^2 + (y-mct)^2 + (z-nct)^2 = R^2.$$

We may therefore say that this surface permanently encloses the same light complex. We inquire as to the quantity of energy enclosed by this surface, viewed in system k, that is, as to the energy of the light complex relatively to the system k.

The spherical surface—viewed in the moving system—is an ellipsoidal surface, the equation for which, at the time $\tau = 0$, is

$$(\beta\xi - l\beta\xi v/c)^2 + (\eta - m\beta\xi v/c)^2 + (\zeta - n\beta\xi v/c)^2 = R^2.$$

If S is the volume of the sphere, and S' that of this ellipsoid, then by a simple calculation

$$\frac{S'}{S} = \frac{\sqrt{(1-v^2/c^2)}}{1-\cos\phi \cdot v/c}.$$

Thus, if we call the light energy enclosed by this surface E when it is measured in the stationary system, and E' when measured in the moving system, we obtain

$$\frac{E'}{E} = \frac{A'^2 S'}{A^2 S} = \frac{1-\cos\phi \cdot v/c}{\sqrt{(1-v^2/c^2)}},$$

and this formula, when $\phi = 0$, simplifies into

$$\frac{E'}{E} = \sqrt{\frac{1-v/c}{1+v/c}}.$$

It is remarkable that the energy and the frequency of a light complex vary with the state of motion of the observer in accordance with the same law.

Now let the co-ordinate plane $\xi = 0$ be a perfectly reflecting surface, at which the plane waves considered in § 7 are reflected. We seek for the pressure of light exerted on the reflecting surface, and for the direction, frequency, and intensity of the light after reflexion.

Let the incidental light be defined by the quantities A, $\cos\phi$, v (referred to system K). Viewed from k the corresponding quantities are

$$A' = A \frac{1-\cos\phi \cdot v/c}{\sqrt{(1-v^2/c^2)}},$$

$$\cos\phi' = \frac{\cos\phi - v/c}{1-\cos\phi \cdot v/c},$$

$$v' = v \frac{1-\cos\phi \cdot v/c}{\sqrt{(1-v^2/c^2)}}.$$

For the reflected light, referring the process to system k, we obtain

$$A'' = A'.$$
$$\cos\phi'' = -\cos\phi'$$
$$v'' = v'.$$

Finally, by transforming back to the stationary system K, we obtain for the reflected light

$$A''' = A'' \frac{1+\cos\phi''.v/c}{\sqrt{(1-v^2/c^2)}} = A\frac{1-2\cos\phi.v/c+v^2/c^2}{1-v^2/c^2},$$

$$\cos\varphi''' = \frac{\cos\phi''+v/c}{1+\cos\phi''.v/c} = -\frac{(1+v^2/c^2)\cos\phi-2v/c}{1-2\cos\phi.v/c+v^2/c^2},$$

$$v''' = v''\frac{1+\cos\phi''v/c}{\sqrt{(1-v^2/c^2)}} = v\frac{1-2\cos\phi.v/c+v^2/c^2}{1-v^2/c^2}.$$

The energy (measured in the stationary system) which is incident upon unit area of the mirror in unit time is evidently $A^2(c\cos\phi-v)/8\pi$. The energy leaving the unit of surface of the mirror in the unit of time is $A'''^2(-c\cos\phi'''+v)/8\pi$. The difference of these two expressions is, by the principle of energy, the work done by the pressure of light in the unit of time. If we set down this work as equal to the product Pv, where P is the pressure of light, we obtain

$$P = 2\cdot\frac{A^2}{8\pi}\frac{(\cos\phi-v/c)^2}{1-v^2/c^2}.$$

In agreement with experiment and with other theories, we obtain to a first approximation

$$P = 2\cdot\frac{A^2}{8\pi}\cos^2\phi.$$

All problems in the optics of moving bodies can be solved by the method here employed. What is essential is, that the electric and magnetic force of the light which is influenced by a moving body, be transformed into a system of co-ordinates at rest relatively to the body. By this means all problems in the optics of moving bodies will be reduced to a series of problems in the optics of stationary bodies.

§ 9. Transformation of the Maxwell–Hertz Equations when Convection-Currents are Taken into Account

We start from the equations

$$\frac{1}{c}\left\{\frac{\partial X}{\partial t}+u_x\varrho\right\} = \frac{\partial N}{\partial y}-\frac{\partial M}{\partial z}, \quad \frac{1}{c}\frac{\partial L}{\partial t}=\frac{\partial Y}{\partial z}-\frac{\partial Z}{\partial y},$$

$$\frac{1}{c}\left\{\frac{\partial Y}{\partial t}+u_y\varrho\right\} = \frac{\partial L}{\partial z}-\frac{\partial N}{\partial x}, \quad \frac{1}{c}\frac{\partial M}{\partial t}=\frac{\partial Z}{\partial x}-\frac{\partial X}{\partial z},$$

$$\frac{1}{c}\left\{\frac{\partial Z}{\partial t}+u_z\varrho\right\} = \frac{\partial M}{\partial x}-\frac{\partial L}{\partial y}, \quad \frac{1}{c}\frac{\partial N}{\partial t}=\frac{\partial X}{\partial y}-\frac{\partial Y}{\partial x},$$

where

$$\varrho = \frac{\partial X}{\partial x}+\frac{\partial Y}{\partial y}+\frac{\partial Z}{\partial z}$$

denotes 4π times the density of electricity, and (u_x, u_y, u_z) the velocity-vector of the charge. If we imagine the electric charges to be invariably coupled to small rigid bodies (ions, electrons), these equations are the electromagnetic basis of the Lorentzian electrodynamics and optics of moving bodies.

Let these equations be valid in the system K, and transform them, with the assistance of the equations of transformation given in §§ 3 and 6, to the system k. We then obtain the equations

$$\frac{1}{c}\left\{\frac{\partial X'}{\partial \tau}+u_\xi\varrho'\right\} = \frac{\partial N'}{\partial \eta}-\frac{\partial M'}{\partial \zeta}, \quad \frac{1}{c}\frac{\partial L'}{\partial \tau}=\frac{\partial Y'}{\partial \zeta}-\frac{\partial Z'}{\partial \eta},$$

$$\frac{1}{c}\left\{\frac{\partial Y'}{\partial \tau}+u_\eta\varrho'\right\} = \frac{\partial L'}{\partial \zeta}-\frac{\partial N'}{\partial \xi}, \quad \frac{1}{c}\frac{\partial M'}{\partial \tau}=\frac{\partial Z'}{\partial \xi}-\frac{\partial X'}{\partial \zeta},$$

$$\frac{1}{c}\left\{\frac{\partial Z'}{\partial \tau}+u_\zeta\varrho'\right\} = \frac{\partial M'}{\partial \xi}-\frac{\partial L'}{\partial \eta}, \quad \frac{1}{c}\frac{\partial N'}{\partial \tau}=\frac{\partial X'}{\partial \eta}-\frac{\partial Y'}{\partial \xi},$$

where

$$u_\xi = \frac{u_x - v}{1 - u_x v/c^2},$$

$$u_\eta = \frac{u_y}{\beta(1 - u_x v/c^2)},$$

$$u_\zeta = \frac{u_z}{\beta(1 - u_x v/c^2)},$$

and

$$\varrho' = \frac{\partial X'}{\partial \xi} + \frac{\partial Y'}{\partial \eta} + \frac{\partial Z'}{\partial \zeta} = \beta(1 - u_x v/c^2)\varrho.$$

Since—as follows from the theorem of addition of velocities (§ 5)—the vector (u_ξ, u_η, u_ζ) is nothing else than the velocity of the electric charge, measured in the system k, we have the proof that, on the basis of our kinematical principles, the electrodynamic foundation of Lorentz's theory of the electrodynamics of moving bodies is in agreement with the principle of relativity.

In addition I may briefly remark that the following important law may easily be deduced from the developed equations: If an electrically charged body is in motion anywhere in space without altering its charge when regarded from a system of co-ordinates moving with the body, its charge also remains—when regarded from the "stationary" system K—constant.

§ 10. Dynamics of the Slowly Accelerated Electron

Let there be in motion in an electromagnetic field an electrically charged particle of charge ε (in the sequel called an "electron"), for the law of motion of which we assume as follows:

If the electron is at rest at a given epoch, the motion of the electron ensues in the next instant of time according to the equations

$$m\frac{d^2x}{dt^2} = \varepsilon X,$$

$$m\frac{d^2y}{dt^2} = \varepsilon Y,$$

$$m\frac{d^2z}{dt^2} = \varepsilon Z$$

where x, y, z denote the co-ordinates of the electron, and m the mass of the electron, as long as its motion is slow.

Now, secondly, let the velocity of the electron at a given epoch be v. We seek the law of motion of the electron in the immediately ensuing instants of time.

Without affecting the general character of our considerations, we may and will assume that the electron, at the moment when we give it our attention, is at the origin of the co-ordinates, and moves with the velocity v along the axis of X of the system K. It is then clear that at the given moment ($t = 0$) the electron is at rest relatively to a system of co-ordinates k which is in parallel motion with constant velocity v along the axis of X.

From the above assumption, in combination with the principle of relativity, it is clear that in the immediately ensuing time (for small values of t) the electron, viewed from the system k, moves in accordance with the equations

$$m\frac{d^2\xi}{d\tau^2} = \varepsilon X',$$

$$m\frac{d^2\eta}{d\tau^2} = \varepsilon Y',$$

$$m\frac{d^2\zeta}{d\tau^2} = \varepsilon Z',$$

in which the symbols ξ, η, ζ, τ, X', Y', Z' refer to the system k. If, further, we decide that when $t = x = y = z = 0$ then $\tau = \xi =$

$\eta = \zeta = 0$, the transformation equations of §§ 3 and 6 hold good, so that we have

$$\xi = \beta(x-vt), \quad \eta = y, \quad \zeta = z, \quad \tau = \beta(t-vx/c^2)$$
$$X' = X, \quad Y' = \beta(Y-vN/c), \quad Z' = \beta(Z+vM/c).$$

With the help of these equations we transform the above equations of motion from system k to system K, and obtain

$$\left.\begin{aligned}\frac{d^2x}{dt^2} &= \frac{\varepsilon}{m\beta^3} X \\ \frac{d^2y}{dt^2} &= \frac{\varepsilon}{m\beta}\left(Y-\frac{v}{c}N\right) \\ \frac{d^2z}{dt^2} &= \frac{\varepsilon}{m\beta}\left(Z+\frac{v}{c}M\right)\end{aligned}\right\}. \tag{A}$$

Taking the ordinary point of view we now inquire as to the "longitudinal" and the "transverse" mass of the moving electron. We write the equations (A) in the form

$$m\beta^3 \frac{d^2x}{dt^2} = \varepsilon X = \varepsilon X',$$
$$m\beta^2 \frac{d^2y}{dt^2} = \varepsilon\beta\left(Y-\frac{v}{c}N\right) = \varepsilon Y',$$
$$m\beta^2 \frac{d^2z}{dt^2} = \varepsilon\beta\left(Z+\frac{v}{c}M\right) = \varepsilon Z',$$

and **remark** firstly that $\varepsilon X'$, $\varepsilon Y'$, $\varepsilon Z'$ are the components of the ponderomotive force acting upon the electron, and are so indeed as viewed in a system moving at the moment with the electron, with the same velocity as the electron. (This force might be measured, for example, by a spring balance at rest in the last-mentioned system.) Now if we call this force simply "the force acting upon the electron,"* and maintain the equation—mass × accelera-

* The definition of force here given is not advantageous, as was first shown by M. Planck. It is more to the point to define force in such a way that the laws of momentum and energy assume the simplest form.

tion = force—and if we also decide that the accelerations are to be measured in the stationary system K, we derive from the above equations

$$\text{Longitudinal mass} = \frac{m}{(\sqrt{1-v^2/c^2})^3}.$$

$$\text{Transverse mass} = \frac{m}{1-v^2/c^2}.$$

With a different definition of force and acceleration we should naturally obtain other values for the masses. This shows us that in comparing different theories of the motion of the electron we must proceed very cautiously.

We remark that these results as to the mass are also valid for ponderable material points, because a ponderable material point can be made into an electron (in our sense of the word) by the addition of an electric charge, *no matter how small*.

We will now determine the kinetic energy of the electron. If an electron moves from rest at the origin of co-ordinates of the system K along the axis of X under the action of an electrostatic force X, it is clear that the energy withdrawn from the electrostatic field has the value $\int \varepsilon X \, dx$. As the electron is to be slowly accelerated, and consequently may not give off any energy in the form of radiation, the energy withdrawn from the electrostatic field must be put down as equal to the energy of motion W of the electron. Bearing in mind that during the whole process of motion which we are considering, the first of the equations (A) applies, we therefore obtain

$$W = \int \varepsilon X \, dx = m \int_0^v \beta^3 v \, dv = mc^2 \left\{ \frac{1}{\sqrt{1-v^2/c^2}} - 1 \right\}.$$

Thus, when $v = c$, W becomes infinite. Velocities greater than that of light have—as in our previous results—no possibility of existence.

This expression for the kinetic energy must also, by virtue of the argument stated above, apply to ponderable masses as well.

We will now enumerate the properties of the motion of the electron which result from the system of equations (A), and are accessible to experiment.

1. From the second equation of the system (A) it follows that an electric force Y and a magnetic force N have an equally strong deflective action on an electron moving with the velocity v, when $Y = Nv/c$. Thus we see that it is possible by our theory to determine the velocity of the electron from the ratio of the magnetic power of deflexion A_m to the electric power of deflexion A_e, for any velocity, by applying the law

$$\frac{A_m}{A_e} = \frac{v}{c}.$$

This relationship may be tested experimentally, since the velocity of the electron can be directly measured, e.g. by means of rapidly oscillating electric and magnetic fields.

2. From the deduction for the kinetic energy of the electron it follows that between the potential difference, P, traversed and the acquired velocity v of the electron there must be the relationship

$$P = \int X \, dx = \frac{m}{\varepsilon} c^2 \left\{ \frac{1}{\sqrt{(1-v^2/c^2)}} - 1 \right\}$$

3. We calculate the radius of curvature R of the path of the electron when a magnetic force N is present (as the only deflective force), acting perpendicularly to the velocity of the electron. From the second of the equations (A) we obtain

$$-\frac{d^2y}{dt^2} = \frac{v^2}{R} = \frac{\varepsilon}{m} \frac{v}{c} N \sqrt{\left(1-\frac{v^2}{c^2}\right)}$$

or

$$R = \frac{mc^2}{\varepsilon} \cdot \frac{v/c}{\sqrt{(1-v^2/c^2)}} \cdot \frac{1}{N}.$$

These three relationships are a complete expression for the laws according to which, by the theory here advanced, the electron must move.

In conclusion I wish to say that in working at the problem here dealt with I have had the loyal assistance of my friend and colleague M. Besso, and that I am indebted to him for several valuable suggestions.

NOTES ON EXTRACT 6

THE greater part of this paper is taken up with a description of the difficulties in reducing the experimental errors sufficiently. The effect to be measured corresponds to $1-(1/\mu\varkappa)$, and the authors compare this prediction of special relativity with a figure quoted from Lorentz and Larmor of $1-(1/\varkappa)$. It is to be noted, however, that before the advent of special relativity, classical electrodynamics was actually unable to calculate the results of such an experiment without *ad hoc* assumptions. Be that as it may, the two quantities quoted were, for the sample, 0·94 and 0·83, the observed effect being 0·96, in an experiment very difficult to perform to high accuracy.

EXTRACT 6[*]

On the Electric Effect of Rotating a Magnetic Insulator in a Magnetic Field

MARJORIE WILSON, B.A., M.SC., and
H. A. WILSON, D.SC., F.R.S.,
Professor of Physics, Rice Institute, Houston, Texas, U.S.A.
(Received May 26—Read June 19, 1913.)

IN a previous paper* by one of us it was shown that when an insulator of specific inductive capacity K rotates in a magnetic field there is an electromotive force induced in it equal to that in a conductor multiplied by $1-K^{-1}$. The object of the experiments described below was to measure the induced electromotive force in a magnetic insulator rotating in a magnetic field parallel to the axis of rotation.

According to the theory based on the "principle of relativity" this induced electromotive force should be equal to that in a conductor multiplied by $1-(\mu K)^{-1}$, where μ is the magnetic permeability of the insulator, whereas on the theory of H. A. Lorentz and Larmor the appropriate multiplier appears to be $1-K^{-1}$, as for a non-magnetic insulator.[†]

No insulator is known for which μ differs appreciably from unity, so that it was necessary to construct a sort of model of a magnetic insulator. The insulator adopted consisted of wax, in which a large number of small steel spheres was embedded. The

[* *Proc. Roy. Soc.* (A), **89**, 99 (1913).]
* "On the Electric Effect of Rotating a Dielectric in a Magnetic Field," by H. A. Wilson, *Phil. Trans.*, 1904. A, vol. 204.
† M. Abraham, *Theorie der Elektrizität*, vol. 11, p. 322.

spheres were $\frac{1}{8}$ inch in diameter, and each one was coated thinly with sealing wax. The coated spheres were packed tightly and melted paraffin wax poured into the empty spaces between them so as to form a solid mass.

The insulator was in the form of a hollow cylinder with inside and outside metal coatings, and was rotated in a magnetic field parallel to the axis of the cylinder. The outside coating was connected to one pair of quadrants of a quadrant electrometer, the other quadrants of which were earthed. If an electromotive force E is induced in the cylinder and this raises the potential of the outer coating by V volts, then

$$E = V(C+C')/C,$$

where C is the capacity between the inner and outer coatings, and C' the capacity of the electrometer connecting wire and outside surface of the outer coating. The inner coating is supposed earthed.

If the potential of the inner coating is raised by an amount E', and this raises the potential of the outer coating by V', then

$$E' = V'(C+C')/C.$$

Let the electrometer deflection due to V be d and that due to V' be d'. Then we have

$$\frac{E}{E'} = \frac{V}{V'} = \frac{d}{d'}.$$

Thus E can be found in terms of d, d', and E'. Another way is to earth the outer coating and charge the inner one to a potential E', then insulate the outer coating and afterwards earth the inner one. In this way a charge $-CE'$ is given to the outer coating which raises its potential to V' and

$$-CE' = V'(C+C').$$

This second method has the disadvantage that it requires the outer coating to be earthed, so that errors may arise due to electric effects produced by opening the key connecting the outer coating to the earth. Also in the first method the potential E' acts in exactly the same way as the induced electromotive force, so that errors due to bad insulation affect both V and V' equally, and so are eliminated. The first method was therefore adopted. In the earlier experiments referred to above a small standard condenser was used to give a known charge to the outer coating, and the capacity of the cylinder was found in terms of that of the condenser. This method was less direct than that now employed.

The inner coating was connected to earth through a 10-ohm resistance which could be connected through a 90-ohm resistance to a dry cell. The potential difference between the ends of the 100 ohms was measured with a Weston voltmeter. When the cell was not connected the inner coating was earthed, and when it was connected the inner coating was raised to a potential one-tenth of that indicated by the voltmeter, which was usually 1·450 volts.

The apparatus was that used in the earlier investigation, with some improvements in detail. The cylinder was 3·73 cm. external diameter, 2 cm. internal diameter, and 9·5 cm. long. The outside surface of the cylinder was covered with a brass tube 0·6 mm. thick, and another brass tube fitted the inside surface. The inner tube was mounted on a shaft, from which it was insulated, and the shaft was mounted between fixed conical bearings and could be rotated by means of a belt driven by a $\frac{1}{2}$-H.P. motor. The cylinder was surrounded by a large solenoid which produced a magnetic field parallel to the axis of rotation.

Two small brass wire brushes made contact, one on the middle of the outer coating and the other on the inner tube close to one end of the cylinder. The arrangement of the brushes is shown in fig. 1. Each brush was kept pressed down lightly but steadily by the weight of a brass bar fastened at right angles to the end of the rod supporting the brush as shown. The inertia of these bars prevented the brushes from jumping when the cylinder was rotating quickly and they could be adjusted so that the pressure on the brushes was very small. This new arrangement of the brushes

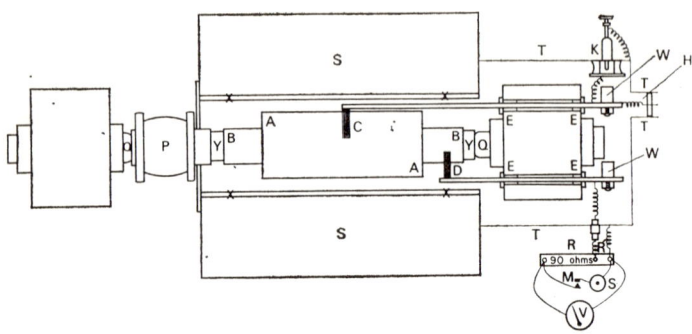

Fig. 1

AA.	Outer coating of cylinder.	WW.	Bars on brush rods.
BB.	Inner coating of cylinder.	K.	Key for earthing outer coating.
P.	Driving pulley.	TTTT.	Metallic screen.
SS.	Solenoid.	H.	Wire leading to electrometer.
C.	Brush on outer coating.	R, R'.	Resistances of 90 to 10 ohms.
D.	Brush on inner coating.	M.	Key.
QQ.	Conical bearings.	S.	Dry cell.
XXXX.	Water jacket.	V.	Voltmeter.
EEEE.	Ebonite bushings supporting brush rods.	YY.	Fibre tube insulating BB from shaft.

caused a great improvement in the working of the apparatus. The electrometer and the wire leading to it were completely enclosed in a metal case which, with the solenoid, formed a complete electrostatic screen around the insulated system. The inside of the solenoid was kept cool by means of a water-jacket. The speed of the cylinder was found with a revolution counter driven by a worm gearing.

To make a determination of the induced electromotive force in the cylinder the electrometer deflection due to changing the potential of the inner coating by about 0·145 volt was first observed and then the cylinder started and its speed found. The effect of reversing a known current in the solenoid was next observed. The speed and sensibility were then again measured. The speed of the cylinder always remained constant within the limits of error. The sensibility also remained constant over long periods. The electrometer scale reading was very steady while the cylinder was rotating and the effect of reversing the curent could be easily and exactly observed. In fact with the new arrangement of the brushes no difficulty in making the observations was experienced and the accuracy seemed to be limited only by the smallness of the deflections obtained. The cylinder insulated well.

There was no effect due to running the cylinder in the absence of a magnetic field and no effect due to reversing the current when the cylinder was at rest. The speed and sensibility were not changed by the current in the solenoid. The sensibility was the same when the cylinder was running as when it was at rest.

The following table contains a set of results obtained:

Revolutions per second (n)	Electrometer sensibility (scale divisions per volt)	Current reversed in ampères (c)	Deflection	Effect in volts (E)	$E/(nc)$
99·0	184	7·5	5·63	0·0306	$4·16 \times 10^{-5}$
98·5	184	15·0	11·4	0·0620	4·20
98·5	184·5	10·0	7·6	0·0412	4·18
198·0	185	14·13	21·5	0·1160	4·14
				Mean...	$4·17 \times 10^{-5}$

It will be seen that the induced electromotive force is very nearly proportional to the current reversed and to the number of revolutions per second. After these observations were made it was found that running the cylinder had produced a narrow air gap between the inner tube and the wax. This was filled up by slightly warming the cylinder and forcing down the mixture of wax and balls at the ends of the cylinder so that the mixture was very tightly pressed against the inner and outer tubes.

The best results could be obtained at about 100 revolutions per second. At greater speeds the electrometer reading was not always quite steady and at smaller speeds the deflections were too small. A set of six concordant measurements of the effect due to reversing about 14 ampères at 100 revolutions per second was therefore obtained and the mean of these was adopted as representing the result of the experiments. The mean results were as follows:

Revolutions per second (n)	Electrometer sensibility	Current reversed (c)	Deflection	Induced E.M.F. (E)	$E/(nc)$
104·0	250	14·34	15·4	0·0616	$4·13 \times 10^{-5}$

It will be seen that this result agrees closely with the others.

In order to compare the observed effect with that in a coductor, it is necessary to know the magnetic induction through the cylinder. Corrections for the induced electromotive forces in the metal coatings have also to be applied.

The change of average induction through the cylinder and its outer coating, due to reversing a current in the solenoid, was found by means of a spiral of fine wire wrapped round it uniformly from end to end. This spiral was connected to a ballistic galvanometer, and the secondary coil of an accurately known mutual induction was included in the circuit. The current in the primary of the mutual induction was measured with the same ammeter that was used to measure the current in the solenoid. The induction was found to be proportional to the current from 5 to 15 ampéres, and to be equal to 4210 per ampère reversed. A second determination of this quantity was done, using a coil of two turns round the cylinder. The induction through this coil was found for a series of equidistant positions along the cylinder and the mean induction through the cylinder calculated. The result was 4200. The mean of the two results, 4205, was adopted. The difference between the mean area of the windings and the area of cross-section of the cylinder was, of course, allowed for. The field at the windings was taken equal to that due to the solenoid in the absence of the cylinder.

The mean induction in the hole through the cylinder was found with a coil which could be slid along inside when the cylinder was supported in its usual position with the inner tube and shaft removed. This was found to be equal to 569 per ampère reversed. The field at the centre of the solenoid in the absence of the cylinder was found to be equal to 205 for a current of 2 ampères, which is the same as the value found in the earlier investigation done in the Cavendish Laboratory.

It was found that near the ends of the cylinder there was a stronger field than inside the hole through it. This, of course, was due to the magnetisation of the cylinder. In consequence of this there was an induced electromotive force in the inner tube which diminished the effect observed. The induction through the cross-section of the inner tube at the brush on it was found to be 740 per ampère reversed, so that the average potential of the inner tube was lowered by the electromotive force due to $740 - 569 = 171$ units of induction.

The induction through the outer cover was taken to be equal per unit cross-section to the field in the absence of the cylinder, which made it 143 per ampère reversed. An error in this quantity would have practically no effect on the final result for the ratio of the effect to that in a conductor, because it is to be subtracted from both quantities.

The induction through the cylinder at the brush on the outer cover was found to be 5010 per ampère reversed. This is greater than the average induction through the cylinder by $5010 - 4205 =$

= 805. In consequence of this the potential at the brush was raised above the average potential of the outside of the cylinder by the potential due to 805 units of induction per ampère reversed.

The observed effect, therefore, includes an induced electromotive force due to $805+143-171 = 777$ units of induction per ampère reversed, acting in the metal coatings of the cylinder. This gives $7 \cdot 77 \times 10^{-6}$ volt per ampère reversed per revolution per second. Subtracting this from the observed effect ($4 \cdot 13 \times 10^{-5}$) we get $3 \cdot 35 \times 10^{-5}$ volt per ampère reversed per revolution per second as the observed effect in the insulator itself.

The average induction through the insulator is $4205-(143+569) = 3493$ per ampère reversed. This would give an induced electromotive force in a conductor equal to $3 \cdot 493 \times 10^{-5}$ volt per ampère reversed per revolution per second. The ratio of the observed effect to that in a conductor is therefore

$$\frac{3 \cdot 35 \times 10^{-5}}{3 \cdot 493 \times 10^{-5}} = 0 \cdot 96.$$

The value of K for the insulator was found* to be $6 \cdot 0$ and that of μ to be $3 \cdot 0$, so that

$$1-(\mu K)^{-1} = 0 \cdot 944, \quad 1-K^{-1} = 0 \cdot 83.$$

The accuracy of the value found for the ratio of the induced electromotive force in the insulator to that in a conductor depends on readings of the voltmeter and ammeter employed. The electromotive force was found in terms of the voltmeter reading, and the induction depends on the product of the mutual induction of the standard and the current determined by the ammeter. The primary of the mutual induction consisted of a single layer of wire wound in a screw thread of 1 mm. pitch cut on a brass tube 5 cm. in diameter and 90 cm. long. The screw was cut on an accurate lathe and the number of threads was found to be 10 per cm. to within 1 part in 5000. The secondary coil consisted of a single layer of 132 turns

* For method see earlier paper referred to above.

wound on an accurately turned brass cylinder, which fitted into the primary coil. The area of the secondary coil was known to within 1 part in 500. The mutual induction was therefore known to a higher order of accuracy than the other quantities involved. The voltmeter and ammeter readings only enter into the final result as the value of the ratio of the potential indicated by the voltmeter to the current indicated by the ammeter, because the same ammeter was used to measure the currents in the solenoid and in the primary of the mutual induction. The value of this ratio was found by means of a standard one-tenth ohm resistance. The following table gives the results obtained:

Current through resistance by ammeter (C)	P.D. between ends of resistance by voltmeter (E)	E/C
5	0·500	0·1
9·95	0·995	0·1
15	1·500	0·1

The ratio E/C is constant and equal to the resistance of the standard, so that it seems certain that no appreciable error could have been introduced by the ammeter and voltmeter, which were new Weston standard instruments. It appears, therefore, that the induced electromotive force in the insulator agrees approximately with that to be expected on the theory of relativity. This theory involves no assumptions as to the constitution of the insulator, so that it is applicable to any medium having in bulk an average permeability μ and an average specific inductive capacity K.

The effect to be expected on the theory of H. A. Lorentz and Larmor depends on assumptions as to the constitution of the material medium so that it is doubtful whether their theory ought to be regarded as necessarily leading to $1 - K^{-1}$ as the value of the ratio of the induced electromotive force in the composite medium employed to that in a conductor.

These experiments therefore confirm the theory of relativity but do not necessarily conflict with the fundamental assumptions of H. A. Lorentz and Larmor's theory. They do, however, make it probable that the application of this theory to magnetic bodies has not yet been worked out in a satisfactory manner.

[*Note added July* 16, 1913.—The specific inductive capacity and permeability of any material medium are average values over volumes large compared with the structural units (molecules or larger bodies) making up the medium. The medium employed has definite values of these quantities for volumes large compared with the volume of one of the spheres used in building it up. It appears, therefore, to be allowable to apply any theoretical results expressed in terms of μ and K to the medium used.]

Our thanks are due to the Government Grant Committee of the Royal Society for a grant with which a large part of the apparatus used in this investigation was originally purchased, and also to the Trustees of the Rice Institute for the facilities for experimental work which they have placed at our disposal.

NOTES ON EXTRACT 7

As WILL be observed in the text, the effect of dispersion on Fresnel's coefficient is different in value for water flowing through a tube from its value for moving glass. It is therefore interesting to note that in this mainly experimental paper, Zeeman is careful to emphasise that his results are for water. The later part of the paper which is omitted deals with further experimental details.

EXTRACT 7[*]

Fresnel's Coefficient for Light of Different Colours (First Part)

By Prof. P. Zeeman

One of the empirical foundations of the electrodynamics of moving bodies in the domain of optics is Fizeau's celebrated experiment on the carrying along of the light waves by the motion of water. Let w be the velocity of water relative to an observer, then for him the velocity of light propagated in the water would be

$$C_a = \frac{c}{\mu} \pm w$$

if the dynamical laws for the addition of velocities were perfectly general.

In this equation μ designs the index of refraction of water, c the velocity of light in vacuo, and we must take the upper or the lower sign, according as the light goes with or against the stream. Fizeau demonstrated that not the entire velocity w but only a fraction of it comes into action. This particular fraction appeared to be approximately equal to $1 - 1/\mu^2$, Fresnel's coefficient. Hence we must write in place of the above given formula:

$$C_a = \frac{c}{\mu} \pm \varepsilon w \tag{1}$$

where

$$\varepsilon = 1 - \frac{1}{\mu^2}. \tag{2}$$

[* *Proc. Acad. Sci. Amst.* **17**, 445 (1914).]

For water ε is equal to seven-sixteenths.

The extremely important role which the formulae (1) and (2) have had in the theory of aberration, in the development of LORENTZ's electronic theory needs not to be exposed here, and it is hardly necessary to state that equation (1) is now regarded as a simple confirmation of EINSTEIN's theorem concerning the addition of velocities.

I may be permitted however to point out the smallness of the second term of formula (1). The velocity which we are able to obtain in a column of water transmitting light is of the order of magnitude of 5 metres per second. We have thus to find a difference of velocity of 5 metres in $3 \times 10^8/(4/3)$ m., i.e. of one part in fifty millions.

This was done by FIZEAU[1] in one of the most ingenious experiments of the whole domain of physics. FIZEAU divided a beam of light issuing from a line of light in the focus of an object-glass into two parallel beams. After traversing two parallel tubes these beams pass through a second lens, in the focus of which a silvered mirror is placed. After reflection the rays are returned to the object glass, interchanging their paths. Each ray thus passes through the two tubes. A system of interference fringes is formed in the focus of the first lens. If water is flowing in opposite directions in the two tubes, one of the interfering beams is always travelling with the current and the other against it. When the water is put in motion a shift of the central white band is observed: by reversing the direction of the current the shift is doubled.

The ingenuity of the arrangement lies in the possibility of securing that the two beams traverse identical ways in opposite directions. Every change due for example to a variation of density or of temperature of the moving medium equally influences the two beams and is therefore automatically compensated.

[1] H. FIZEAU, Sur les hypothèses relatives à l'éther lumineux et sur une expérience qui paraît démontrer que le mouvement des corps change la vitesse avec laquelle la lumière se propage dans leur intérieur. *Ann. de Chim. et de Phys.* (3) **57**, 385. 1859.

One can be sure that a shift of the system of interference fringes, observed when reversing the direction of the current must be due to a change of the velocity of propagation of the light.

The tubes used by FIZEAU had a length of about 1·5 metres and an internal diameter of 5·3 m.m., whereas the velocity of the water was estimated at 7 metres. With *white light* the shift of the central band of the system of interference fringes observed by reversing the direction of flow was found from 19 rather concordant observations equal to 0·46 of the distance of two fringes; the value calculated with FRESNEL'S coefficient is 0·404.

The result is favourable to the theory of FRESNEL. The amount of the shift is less than would correspond to the full velocity of the water and also agrees numerically with a coefficient $1 - 1/\mu^2$, if the uncertainty of the observations is taken into account.

FIZEAU'S experiments, though made by a method which is theoretically as simple as it is perfect, left some doubts as to their accuracy, partly by reason of the remarkable conclusions as to relative motion of ether and matter to which they gave rise, and these doubts could only be removed by new experiments.

35 years after FIZEAU'S first communication[1] to the Académie des Sciences, MICHELSON and MORLEY[2] repeated the experiment. They intended to remove some difficulties inherent to FIZEAU'S method of observation and also, if possible, to measure accurately the fraction to be applied to the velocity of the water. MICHELSON uses the principle of his interferometer and produces interference fringes of considerable width without reducing at the same time the intensity of the light. The arrangement is further the same as that used by FIZEAU but performed with the considerable means, which American scientists have at their disposal for important scientific questions. The internal diameter of the tubes in the experiment of MICHELSON and MORLEY was 28 m.m. and in a first

[1] *Comptes rendus* **33**, 349, 1851.
[2] A. A. MICHELSON and E. W. MORLEY, Influence of motion of the medium on the velocity of light. *Am. Journ. of Science* (3) **31**, 377, 1886.

series the *total*[1] length of the tubes was 3 metres, in a second series a little more than 6 metres.

From three series of experiments with *white light* MICHELSON found results which if reduced to what they would be if the tube were 2×5 metres long and the velocity 1 metre per second, would be as follows:

"Series Δ = double displacement

  1   0·1858
  2   0·1838
  3   0·1800"

"The final weighted value of Δ for all the observations is $\Delta = 0·1840$. From this by substitution in the formula, we get $\varepsilon = 0·434$ with a possible error of $+0·002$".

For light of the wavelength of the *D*-lines we calculate $1 - 1/\mu^2 = 0·437$. This agreement between theory and observation is extremely satisfactory.

A new formula for ε was given by LORENTZ[2] in 1895 viz.:

$$\varepsilon = 1 - \frac{1}{\mu^2} - \frac{1}{\mu} \lambda \frac{d\mu}{d\lambda} \qquad (3)$$

For the wavelength of the sodium lines this becomes:

0·451.

We see, therefore, that the value deduced by formula (3) deviates more from the result of the observations than the value given by the simple formula (2).

"Sollte es gelingen, was zwar schwierig, aber nicht unmöglich scheint, experimentell zwischen den Gleichungen (3) und (2) zu entscheiden, und sollte sich dabei die erstere bewähren, so hätte

---

[1] *Viz.* the sum of the lengths of the ways in the moving medium, traversed by each of the interfering beams, or approximately twice the length of one of the tubes.

[2] H. A. LORENTZ, *Versuch einer Theorie der electrischen und optischen Erscheinungen in bewegten Körpern*, p. 101, 1895. See also *Theory of Electrons*, p. 290.

man gleichsam die DOPPLER'sche Veränderung der Schwingungsdauer für eine künstlich erzeugte Geschwindigkeit beobachtet. Es ist ja nur unter Berücksichtigung dieser Veränderung, dass wir die Gleichung (3) abgeleitet haben".[1]

It seemed of some importance to repeat *with light of different colours* FIZEAU's experiment, now that the correspondence between theory and observation had become less brilliant, and in view of the fundamental importance of the experiment for the optics of moving bodies.

From the point of view of the theory of relativity the formula (3) is easily proved, as has been pointed out by LAUE[2], neglecting terms of the order w^2/c^2. Recently, however, again some doubt as to the exactness of LORENTZ's term has been expressed. I may refer here to a remark by MAX B. WEINSTEIN[3] in a recent publication and to a paper by G. JAUMANN.[4] The last mentioned physicist gives an expression for the coefficient ε, which for water does not differ much, but in other cases deviates very considerably from FRESNEL's coefficient.

The interference fringes were produced by the method of MICHELSON. The method of observation introduced will be described later on. The incident ray $s l a$ meets a slightly silvered plate at a. Here it divides into a reflected and a transmitted part. The reflected ray follows the path $a\,b\,c\,d\,e\,a\,f$, the transmitted one the path $e\,a\,d\,c\,b\,a\,f$. These rays meeting in the focal plane of f have pursued identical, not only equivalent, paths, at least this is the case for that part of the system of interference fringes which in white light forms the centre of the central band.

In order to verify the formula (3) it is necessary that the light

[1] LORENTZ, *Versuch* u.s.w., p. 102.

[2] M. LAUE, Die Mitführung des Lichtes durch bewegte Körper nach dem Relativitätsprinzip. *Ann. d. Phys.* **23**, 989. 1907.

[3] MAX B. WEINSTEIN, *Die Physik der bewegten Materie und die Relativitäts theorie.* Leipzig, 1913, see note on p. 227 of his publication.

[4] G. JAUMANN, Elektromagnetische Theorie. *Sitzungsber. d. Kaiserl. Ak. der Wiss. Wien, mathem. naturw. Kl.* **117**, 379, 1908, especially p. 459.

be monochromatic. Further it seems of immense advantage to have a water current which remains constant during a considerable time.

For observations with violet light this even becomes strictly necessary, because visual observations are impossible with the violet mercury line (4358) used. Michelson obtained a flow of water by filling a tank, connected with the apparatus; by means of large valves the current was made to flow in either direction through the tubes. "The flow lasted about three minutes, which gave time for a number of observations with the flow in alternating directions". In view of my experiments the municipal authorities of Amsterdam permitted the connection of a pipe of 7·5 cm. internal diameter to the main water conduit. There was no difficulty now photographing the violet system of interference fringes, though the time of exposition with one direction of flow was between 5 and 7 minutes. The pressure of the water proved to be very constant during a series of observations; the maximum velocity in the axis of the tubes, of 40 mm. internal diameter and of a total length of 6 metres, was about 5·5 metres.

Before recording some details of my experiments, I may be permitted to communicate the general result that for water *there exists a dispersion of* FRESNEL'S *coefficient* and *that formula* (3) *and therefore the third term of* LORENTZ *is essentially correct*.

I wish to record here my thanks to Mr. W. DE GROOT phil. nat. cand. and assistant in the physical laboratory for his assistance during my experiments with the final apparatus.

The difficulties encountered in these experiments were only surmounted after two reconstructions of the apparatus. Great annoyance gave the inconstancy of the interference fringes, when the pressure of the water or the direction of flow were changed. Then not only the width of the interference bands, but the inclination of the fringes were undergoing uncontrollable variations. All these defects were perfectly eliminated by the use of wide tubes and by arranging the end plates in the manner indicated in Fig. 3.

I am indebted to Mr. J. VAN DER ZWAAL, instrumentmaker in the laboratory for his carefully carrying out my instructions and designs in the mechanical construction of the apparatus.

In Fig. 2A a side aspect, and in Fig. 2B a horizontal projection of the arrangement on a scale of about $\frac{1}{15}$th is given.

The interferometer is at the right side, at the left the rectangular prism is placed.

The mounting of this prism is only sketched and was in reality more stable than might be inferred from the drawing.

Prism and interferometer were mounted on the piers cemented to the large brick pier of the laboratory. The tubes are entirely disconnected from the interferometer and mounted on a large iron I girder; this girder is placed upon piers of freestone cemented to large plates of freestone fixed to the wooden laboratory floors. In this manner the adjustment of the interferometer cannot be disturbed by vibrations proceeding from the tubes. At the right of the horizontal projection the four large valves may be seen, by turning which the current was made to flow in either direction through the tube systems.

The mountings containing the glass plates by which the tubes are closed are not given in the Plate. One of these mountings containing the plane parallel plates of glass is drawn to scale in Fig. 3 at one half of the natural size. The four plates of glass are by Hilger, they are circular of 24 mm. diameter and 10 mm. thick; in a second series of observations plates 7 mm. thick have been used. The accuracy of parallelism of the plates is excellent; they are indeed cut from echelon plates. The general plan adopted for the construction of the plate mountings is this: one can only be sure that no change will occur in the position of the plates during the course of an experiment, if this position is *entirely definite*. In order to attain this the glass plate rests upon the inner, accurately grinded, surface of the brass piece d. This piece d fits accurately into the conical inner part of a piece b, itself rigidly screwed to the tube a. Part d and b are connected by means of the counter nut *c*. The glassplate is held against *d* by the nut *e*. There is no objection to the presence at the *inside* between *e* and *d* of rings of hard india-rubber and of brass.

NOTES ON EXTRACT 8

DIRAC'S approach in this paper is admirably summarised in the first paragraph; to find why nature had chosen, not a point charge, but a spinning electron, as a building brick. He then proceeds to relate this with the necessity of making the theory Lorentz invariant. It is clear now (what was perhaps less clear in 1928) that neither requirement involved the other uniquely; it is not necessary for Lorentz invariance actually to have half-integral spin—other spins are possible and the choice is experimental. Similarly Pauli and Darwin had already fitted a spinning electron, at least in a formal manner into a pre-relativistic theory. Dirac's achievement is to have shown how the two requirements can be combined.

EXTRACT 8[*]

The Quantum Theory of the Electron

By P. A. M. DIRAC

St. John's College, Cambridge

(Communicated by R. H. Fowler, F.R.S.—Received January 2, 1928)

THE new quantum mechanics, when applied to the problem of the structure of the atom with point-charge electrons, does not give results in agreement with experiment. The discrepancies consist of "duplexity" phenomena, the observed number of stationary states for an electron in an atom being twice the number given by the theory. To meet the difficulty, Goudsmit und Uhlenbeck have introduced the idea of an electron with a spin angular momentum of half a quantum and a magnetic moment of one Bohr magneton. This model for the electron has been fitted into the new mechanics by Pauli[*], and Darwin[†], working with an equivalent theory, has shown that it gives results in agreement with experiment for hydrogen-like spectra to the first order of accuracy.

The question remains as to why Nature should have chosen this particular model for the electron instead of being satisfied with the point-charge. One would like to find some incompleteness in the previous methods of applying quantum mechanics to the point-charge electron such that, when removed, the whole of the duplexity phenomena follow without arbitrary assumptions. In the present paper it is shown that this is the case, the incomplete-

[* *Proc. Roy. Soc.* (A), **117**, 610 (1928).]
* Pauli, *Z. f. Physik*, vol. **43**, p. 601 (1927).
† Darwin, *Roy. Soc. Proc.*, A, vol. 116, p. 227 (1927).

ness of the previous theories lying in their disagreement with relativity, or, alternatively, with the general transformation theory of quantum mechanics. It appears that the simplest Hamiltonian for a point-charge electron satisfying the requirements of both relativity and the general transformation theory leads to an explanation of all duplexity phenomena without further assumption. All the same there is a great deal of truth in the spinning electron model, at least as a first approximation. The most important failure of the model seems to be that the magnitude of the resultant orbital angular momentum of an electron moving in an orbit in a central field of force is not a constant, as the model leads one to expect.

§ 1. Previous Relativity Treatments

The relativity Hamiltonian according to the classical theory for a point electron moving in an arbitrary electro-magnetic field with scalar potential A_0 and vector potential A is

$$F \equiv -\left(\frac{W}{c}+\frac{e}{c}A_0\right)^2 + \left(\mathbf{p}+\frac{e}{c}\mathbf{A}\right)^2 + m^2c^2,$$

where $\mathbf{p}$ is the momentum vector. It has been suggested by Gordon[*] that the operator of the wave equation of the quantum theory should be obtained from this F by the same procedure as in non-relativity theory, namely, by putting

$$W = ih\frac{\partial}{\partial t},$$

$$p_r = -ih\frac{\partial}{\partial x_r}, \quad r = 1, 2, 3,$$

[*] Gordon, *Z. f. Physik*, vol. 40, p. 117 (1926).

in it. This gives the wave equation

$$F\psi \equiv \left[-\left(ih\frac{\partial}{c\,\partial t}+\frac{e}{c}A_0\right)^2 + \Sigma_r\left(-ih\frac{\partial}{\partial x_r}+\frac{e}{c}A_r\right)^2 + m^2c^2\right]\psi = 0, \quad (1)$$

the wave function ψ being a function of x_1, x_2, x_3, t. This gives rise to two difficulties.

The first is in connection with the physical interpretation of ψ. Gordon, and also independently Klein,[†] from considerations of the conservation theorems, make the assumption that if ψ_m, ψ_n are two solutions

$$\varrho_{mn} = -\frac{e}{2mc^2}\left\{ih\left(\psi_m\frac{\partial\psi_n}{\partial t}-\bar{\psi}_n\frac{\partial\psi_m}{\partial t}\right)+2eA_0\bar{\psi}_n\right\}$$

and

$$\mathbf{I}_{mn} = -\frac{e}{2m}\left\{-ih(\psi_m\,\text{grad}\,\bar{\psi}_n-\bar{\psi}_n\,\text{grad}\,\psi_m)+2\frac{e}{c}\mathbf{A}_m\psi_m\bar{\psi}_n\right\}$$

are to be interpreted as the charge and current associated with the transition $m \to n$. This appears to be satisfactory so far as emission and absorption of radiation are concerned, but is not so general as the interpretation of the nonrelativity quantum mechanics, which has been developed[‡] sufficiently to enable one to answer the question: What is the probability of any dynamical variable at any specified time having a value lying between any specified limits, when the system is represented by a given wave function ψ_n? The Gordon–Klein interpretation can answer such questions if they refer to the position of the electron (by the use of ϱ_{nn}), but not if they refer to its momentum, or angular momentum or any other dynamical variable. We should expect the interpretation

[†] Klein, *Z. f. Physik*, vol. 41, p. 407 (1927).

[‡] Jordan, *Z. f. Physik*, vol. 40, p. 809 (1927); Dirac, *Roy. Soc. Proc.*, A, vol. 113, p. 621 (1927).

of the relativity theory to be just as general as that of the non-relativity theory.

The general interpretation of non-relativity quantum mechanics is based on the transformation theory, and is made possible by the wave equation being of the form

$$(H-W)\psi = 0, \qquad (2)$$

i.e., being linear in W or $\partial/\partial t$, so that the wave function at any time determines the wave function at any later time. The wave equation of the relativity theory must also be linear in W if the general interpretation is to be possible.

The second difficulty in Gordon's interpretation arises from the fact that if one takes the conjugate imaginary of equation (1), one gets

$$\left[-\left(-\frac{W}{c}+\frac{e}{c}A_0\right)^2+\left(-\mathbf{p}+\frac{e}{c}\mathbf{A}\right)^2+m^2c^2\right]\psi = 0,$$

which is the same as one would get if one put $-e$ for e. The wave equation (1) thus refers equally well to an electron with charge e as to one with charge $-e$. If one considers for definiteness the limiting case of large quantum numbers one would find that some of the solutions of the wave equation are wave packets moving in the way a particle of charge $-e$ would move on the classical theory, while others are wave packets moving in the way a particle of charge e would move classically. For this second class of solutions W has a negative value. One gets over the difficulty on the classical theory by arbitrarily excluding those solutions that have a negative W. One cannot do this on the quantum theory, since in general a perturbation will cause transitions from states with W positive to states with W negative. Such a transition would appear experimentally as the electron suddenly changing its charge from $-e$ to e, a phenomenon which has not been observed. The true relativity wave equation should thus be such that its solutions split up into two non-combining sets, referring respectively to the charge $-e$ and the charge e.

In the present paper we shall be concerned only with the removal of the first of these two difficulties. The resulting theory is therefore still only an approximation, but it appears to be good enough to account for all the duplexity phenomena without arbitrary assumptions.

§ 2. The Hamiltonian for No Field

Our problem is to obtain a wave equation of the form (2) which shall be invariant under a Lorentz transformation and shall be equivalent to (1) in the limit of large quantum numbers. We shall consider first the case of no field, when equation (1) reduces to

$$(-p_0^2 + \mathbf{p}^2 + m^2c^2)\psi = 0 \tag{3}$$

if one puts

$$p_0 = \frac{W}{c} = ih\frac{\partial}{c\,\partial t}.$$

The symmetry between p_0 and p_1, p_2, p_3 required by relativity shows that, since the Hamiltonian we want is linear in p_0, it must also be linear in p_1, p_2 and p_3. Our wave equation is therefore of the form

$$(p_0 + \alpha_1 p_1 + \alpha_2 p_2 + \alpha_3 p_3 + \beta)\psi = 0, \tag{4}$$

where for the present all that is known about the dynamical variables or operators α_1, α_2, α_3, β is that they are independent of p_0, p_1, p_2, p_3, *i.e.*, that they commute with t, x_1, x_2, x_3. Since we are considering the case of a particle moving in empty space, so that all points in space are equivalent, we should expect the Hamiltonian not to involve t, x_1, x_2, x_3. This means that α_1, α_2, α_3, β are independent of t, x_1, x_2, x_3, *i.e.*, that they commute with p_0, p_1, p_2, p_3. We are therefore obliged to have other dynamical variables besides the co-ordinates and momenta of the electron, in order that α_1, α_2, α_3, β may be functions of them. The wave function ψ must then involve more variables than merely x_1, x_2, x_3, t.

Equation (4) leads to

$$0 = (-p_0+\alpha_1 p_1+\alpha_2 p_2+\alpha_3 p_3+\beta)(p_0+\alpha_1 p_1+\alpha_2 p_2+\alpha_3 p_3+\beta)\psi.$$
$$= [-p_0^2+\Sigma\alpha_1^2 p_1^2+\Sigma(\alpha_1\alpha_2+\alpha_2\alpha_1)p_1 p_2+\beta^2+\Sigma(\alpha_1\beta+\beta\alpha_1)p_1]\psi, \quad (5)$$

where the Σ refers to cyclic permutation of the suffixes 1, 2, 3. This agrees with (3) if

$$\left.\begin{array}{l}\alpha_r^2 = 1, \quad \alpha_r\alpha_s+\alpha_s\alpha_r = 0 \quad (r \neq s) \\ \beta^2 = m^2 c^2, \quad \alpha_r\beta+\beta\alpha_r = 0\end{array}\right\} r, s = 1, 2, 3.$$

If we put $\beta = \alpha_4 mc$, these conditions become

$$\alpha_\mu^2 = 1 \quad \alpha_\mu\alpha_\nu+\alpha_\nu\alpha_\mu = 0 \quad (\mu \neq \nu) \quad \mu, \nu = 1, 2, 3, 4. \quad (6)$$

We can suppose the α_μ's to be expressed as matrices in some matrix scheme, the matrix elements of α_μ being, say, $\alpha_\mu(\zeta'\zeta'')$. The wave function ψ must now be a function of ζ as well as x_1, x_2, x_3, t. The result of α_μ multiplied into ψ will be a function $(\alpha_\mu\psi)$ of x_1, x_2, x_3, t, ζ defined by

$$(\alpha_\mu\psi)(x, t, \zeta) = \Sigma_{\zeta'}\alpha_\mu(\zeta\zeta')\psi(x, t, \zeta').$$

We must now find four matrices α_μ to satisfy the conditions (6). We make use of the matrices

$$\sigma_1 = \begin{pmatrix}0 & 1\\1 & 0\end{pmatrix} \quad \sigma_2 = \begin{pmatrix}0 & -i\\i & 0\end{pmatrix} \quad \sigma_3 = \begin{pmatrix}1 & 0\\0 & -1\end{pmatrix}$$

which Pauli introduced[*] to describe the three components of spin angular momentum. These matrices have just the properties

$$\sigma_r^2 = 1 \quad \sigma_r\sigma_s+\sigma_s\sigma_r = 0, \quad (r \neq s), \quad (7)$$

that we require for our α's. We cannot, however, just take the σ's to be three of our α's, because then it would not be possible to find the fourth. We must extend the σ's in a diagonal manner to

[*] Pauli, *loc. cit.*

bring in two more rows and columns, so that we can introduce three more matrices $\varrho_1, \varrho_2, \varrho_3$ of the same form as $\sigma_1, \sigma_2, \sigma_3$, but referring to different rows and columns, thus:

$$\sigma_1 = \begin{Bmatrix} 0 & 1 & 0 & 0 \\ 1 & 0 & 0 & 0 \\ 0 & 0 & 0 & 1 \\ 0 & 0 & 1 & 0 \end{Bmatrix} \quad \sigma_2 = \begin{Bmatrix} 0 & -i & 0 & 0 \\ i & 0 & 0 & 0 \\ 0 & 0 & 0 & -i \\ 0 & 0 & i & 0 \end{Bmatrix}$$

$$\sigma_3 = \begin{Bmatrix} 1 & 0 & 0 & 0 \\ 0 & -1 & 0 & 0 \\ 0 & 0 & 1 & 0 \\ 0 & 0 & 0 & -1 \end{Bmatrix},$$

$$\varrho_1 = \begin{Bmatrix} 0 & 0 & 1 & 0 \\ 0 & 0 & 0 & 1 \\ 1 & 0 & 0 & 0 \\ 0 & 1 & 0 & 0 \end{Bmatrix} \quad \varrho_2 = \begin{Bmatrix} 0 & 0 & -i & 0 \\ 0 & 0 & 0 & -i \\ i & 0 & 0 & 0 \\ 0 & i & 0 & 0 \end{Bmatrix}$$

$$\varrho_3 = \begin{Bmatrix} 1 & 0 & 0 & 0 \\ 0 & 1 & 0 & 0 \\ 0 & 0 & -1 & 0 \\ 0 & 0 & 0 & -1 \end{Bmatrix}.$$

The ϱ's are obtained from the σ's by interchanging the second and third rows, and the second and third columns. We now have, in addition to equations (7)

and also $$\left.\begin{array}{c} \varrho_r^2 = 1 \quad \varrho_r\varrho_s + \varrho_s\varrho_r = 0 \quad (r \neq s). \\ \varrho_r\sigma_t = \sigma_t\varrho_r. \end{array}\right\} \quad (7')$$

If we now take

$$\alpha_1 = \varrho_1\sigma_1, \quad \alpha_2 = \varrho_1\sigma_2, \quad \alpha_3 = \varrho_1\sigma_3, \quad \alpha_4 = \varrho_3,$$

all the conditions (6) are satisfied, *e.g.*,

$$\alpha_1^2 = \varrho_1\sigma_1\varrho_1\sigma_1 = \varrho_1^2\sigma_1^2 = 1$$
$$\alpha_1\alpha_2 = \varrho_1\sigma_1\varrho_1\sigma_2 = \varrho_1^2\sigma_1\sigma_2 = -\varrho_1^2\sigma_2\sigma_1 = -\alpha_2\alpha_1.$$

The following equations are to be noted for later reference

$$\left.\begin{array}{l}\varrho_1\varrho_2 = i\varrho_3 = -\varrho_2\varrho_1 \\ \sigma_1\sigma_2 = i\sigma_3 = -\sigma_2\sigma_1\end{array}\right\}, \quad (8)$$

together with the equations obtained by cyclic permutation of the suffixes.

The wave equation (4) now takes the form

$$[p_0 + \varrho_1(\boldsymbol{\sigma}, \mathbf{p}) + \varrho_3 mc]\psi = 0, \quad (9)$$

where σ denotes the vector $(\sigma_1, \sigma_2, \sigma_3)$. $((\sigma, p) \equiv \Sigma \sigma_r p_r$, $\sigma_r p_r$ being a matrix product).

§ 3. Proof of Invariance under a Lorentz Transformation

Multiply equation (9) by ϱ_3 on the left-hand side. It becomes, with the help of (8),

$$[\varrho_3 p_0 + i\varrho_2(\sigma_1 p_1 + \sigma_2 p_2 + \sigma_3 p_3) + mc]\psi = 0.$$

Putting

$$p_0 = ip_4,$$

$$\varrho_3 = \gamma_4, \quad \varrho_2 \sigma_r = \gamma_r, \quad r = 1, 2, 3, \quad (10)$$

we have $\quad [i\Sigma\gamma_\mu p_\mu + mc]\psi = 0, \quad \mu = 1, 2, 3, 4. \quad (11)$

The p_μ transform under a Lorentz transformation according to the law

$$p'_\mu = \Sigma_\nu a_{\mu\nu} p_\nu,$$

where the coefficients $a_{\mu\nu}$ are c-numbers satisfying

$$\Sigma_\mu a_{\mu\nu} a_{\mu\tau} = \delta_{\nu\tau}, \quad \Sigma_\tau a_{\mu\tau} a_{\nu\tau} = \delta_{\mu\nu}.$$

The wave equation therefore transforms into

$$[i\Sigma\gamma'_\mu p'_\mu + mc]\psi = 0, \quad (12)$$

where

$$\gamma'_\mu = \Sigma_\nu a_{\mu\nu} \gamma_\nu.$$

Now the γ_μ, like the α_μ, satisfy

$$\gamma_\mu^2 = 1, \quad \gamma_\mu\gamma_\nu + \gamma_\nu\gamma_\mu = 0, \quad (\mu \neq \nu).$$

These relations can be summed up in the single equation

$$\gamma_\mu\gamma_\nu + \gamma_\nu\gamma_\mu = 2\delta_{\mu\nu}.$$

We have

$$\begin{aligned}\gamma'_\mu\gamma'_\nu + \gamma'_\nu\gamma'_\mu &= \Sigma_{\tau\lambda}a_{\mu\tau}a_{\nu\lambda}(\gamma_\tau\gamma_\lambda + \gamma_\lambda\gamma_\tau) \\ &= 2\Sigma_{\tau\lambda}a_{\mu\tau}a_{\nu\lambda}\delta_{\tau\lambda} \\ &= 2\Sigma_\tau a_{\mu\tau}a_{\nu\tau} = 2\delta_{\mu\nu}.\end{aligned}$$

Thus the γ'_μ satisfy the same relations as the γ_μ. Thus we can put, analogously to (10)

$$\gamma'_4 = \varrho'_3, \quad \gamma'_r = \varrho'_2\sigma'_r$$

where the ϱ''s and σ''s are easily verified to satisfy the relations corresponding to (7), (7') and (8), if ϱ'_2 and ϱ'_1 are defined by $\varrho'_2 = -i\gamma'_1\gamma'_2\gamma'_3$, $\varrho'_1 = -i\varrho'_2\varrho'_3$.

We shall now show that, by a canonical transformation, the ϱ''s and σ''s may be brought into the form of the ϱ's and σ's. From the equation $\varrho'^2_3 = 1$, it follows that the only possible characteristic values for ϱ'_3 are ± 1. If one applies to ϱ'_3 a canonical transformation with the transformation function ϱ'_1, the result is

$$\varrho'_1\varrho'_3(\varrho'_1)^{-1} = -\varrho'_3\varrho'_1(\varrho'_1)^{-1} = -\varrho'_3.$$

Since characteristic values are not changed by a canonical transformation, ϱ'_3 must have the same characteristic values as $-\varrho'_3$. Hence the characteristic values of ϱ'_3 are $+1$ twice and -1 twice. The same argument applies to each of the other ϱ''s, and to each of the σ''s.

Since ϱ'_3 and σ'_3 commute, they can be brought simultaneously to the diagonal form by a canonical transformation. They will then have for their diagonal elements each $+1$ twice and -1 twice. Thus, by suitably rearranging the rows and columns, they can be brought into the form ϱ_3 and ϱ_3 respectively. (The possibility $\varrho'_3 = \pm\sigma'_3$ is excluded by the existence of matrices that commute with one but not with the other.)

Any matrix containing four rows and columns can be expressed as
$$c+\Sigma_r c_r \sigma_r+\Sigma_r c'_r \varrho_r+\Sigma_{rs} c_{rs} \varrho_r \sigma_s \tag{13}$$

where the sixteen coefficients c, c_r, c'_r, c_{rs} are c-numbers. By expressing σ'_1 in this way, we see, from the fact that it commutes with $\varrho'_3 = \varrho_3$ and anticommutes* with $\sigma'_3 = \sigma_3$, that it must be of the form

$$\sigma'_1 = c_1\sigma_1 + c_2\sigma_2 + c_{31}\varrho_3\sigma_1 + c_{32}\varrho_3\sigma_2,$$

i.e., of the form

$$\sigma'_1 = \begin{Bmatrix} 0 & a_{12} & 0 & 0 \\ a_{21} & 0 & 0 & 0 \\ 0 & 0 & 0 & a_{34} \\ 0 & 0 & a_{43} & 0 \end{Bmatrix}$$

The condition $\sigma'^2_1 = 1$ shows that $a_{12}a_{21} = 1$, $a_{34}a_{43} = 1$. If we now apply the canonical transformation: first row to be multiplied by $(a_{21}/a_{12})^{1/2}$ and third row to be multiplied by $(a_{43}/a_{34})^{1/2}$, and first and third columns to be divided by the same expressions, σ'_1 will be brought into the form of σ_1, and the diagonal matrices σ'_3 and ϱ'_3 will not be changed.

If we now express ϱ'_1 in the form (13) and use the conditions that it commutes with $\sigma'_1 = \sigma_1$ and $\sigma'_3 = \sigma_3$ and anticommutes with $\varrho'_3 = \varrho_3$, we see that it must be of the form

$$\varrho'_1 = c'_1\varrho_1 + c'_2\varrho_2.$$

The condition $\varrho'^2_1 = 1$ shows that $c'^2_1 + c'^2_2 = 1$, or $c'_1 = \cos\theta$, $c'_2 = \sin\theta$. Hence ϱ'_1 is of the form

$$\varrho'_1 = \begin{Bmatrix} 0 & 0 & e^{-i\theta} & 0 \\ 0 & 0 & 0 & e^{-i\theta} \\ e^{i\theta} & 0 & 0 & 0 \\ 0 & e^{i\theta} & 0 & 0 \end{Bmatrix}$$

* We say that a anticommutes with b when $ab = -ba$.

If we now apply the canonical transformation: first and second rows to be multiplied by $e^{i\theta}$ and first and second columns to be divided by the same expression, ϱ_1' will be brought into the form ϱ_1, and σ_1, σ_3, ϱ_3 will not be altered. ϱ_2' and σ_2' must now be of the form ϱ_2 and σ_2, on account of the relations $i\varrho_2' = \varrho_3'\varrho_1'$, $i\sigma_2' = \sigma_3'\sigma_1'$.

Thus by a succession of canonical transformations, which can be combined to form a single canonical transformation, the ϱ''s and σ''s can be brought into the form of the ϱ's and σ's. The new wave equation (12) can in this way be brought back into the form of the original wave equation (11) or (9); so that the results that follow from this original wave equation must be independent of the frame of reference used.

§ 4. The Hamiltonian for an Arbitrary Field

To obtain the Hamiltonian for an electron in an electromagnetic field with scalar potential A_0 and vector potential $\mathbf{A}$, we adopt the usual procedure of substituting $p_0 + e/c \cdot A_0$ for p_0 and $\mathbf{p} + e/c \cdot \mathbf{A}_0$ for $\mathbf{p}$ in the Hamiltonian for no field. From equation (9) we thus obtain

$$\left[p_0 + \frac{e}{c} A_0 + \varrho_1 \left(\boldsymbol{\sigma}, \mathbf{p} + \frac{e}{c} \mathbf{A} \right) + \varrho_3 mc \right] \psi = 0. \qquad (14)$$

This wave equation appears to be sufficient to account for all the duplexity phenomena. On account of the matrices ϱ and σ containing four rows and columns, it will have four times as many solutions as the non-relativity wave equation, and twice as many as the previous relativity wave equation (1). Since half the solutions must be rejected as referring to the charge $+e$ on the electron, the correct number will be left to account for duplexity phenomena. The proof given in the preceding section of invariance under a Lorentz transformation applies equally well to the more general wave equation (14).

248   SPECIAL RELATIVITY

We can obtain a rough idea of how (14) differs from the previous relativity wave equation (1) by multiplying it up analogously to (5). This gives, if we write e' for e/c,

$$\begin{aligned}
0 &= [-(p_0+e'A_0)+\varrho_1(\boldsymbol{\sigma},\mathbf{p}+e'\mathbf{A})+\varrho_3 mc] \\
&\quad \times [(p_0+e'A_0)+\varrho_1(\boldsymbol{\sigma},\mathbf{p}+e'\mathbf{A})+\varrho_3 mc]\psi \\
&= [-(p_0+e'A_0)^2+(\boldsymbol{\sigma},\mathbf{p}+e'\mathbf{A})^2+m^2c^2 \\
&\quad +\varrho_1\{(\boldsymbol{\sigma},\mathbf{p}+e'\mathbf{A})(p_0+e'A_0)-(p_0+e'A_0)(\boldsymbol{\sigma},\mathbf{p}+e'\mathbf{A})\}]\psi.
\end{aligned} \quad (15)$$

We now use the general formula, that if **B** and **C** are any two vectors that commute with **σ**

$$\begin{aligned}
(\boldsymbol{\sigma},\mathbf{B})(\boldsymbol{\sigma},\mathbf{C}) &= \Sigma\sigma_1^2 B_1 C_1 + \Sigma(\sigma_1\sigma_2 B_1 C_2 + \sigma_2\sigma_1 B_2 C_1) \\
&= (\mathbf{B},\mathbf{C})+i\Sigma\sigma_3(B_1 C_2-B_2 C_1) \\
&= (\mathbf{B},\mathbf{C})+i(\boldsymbol{\sigma},\mathbf{B}\times\mathbf{C}).
\end{aligned} \quad (16)$$

Taking $\mathbf{B} = \mathbf{C} = \mathbf{p}+e'\mathbf{A}$, we find

$$\begin{aligned}
(\boldsymbol{\sigma},\mathbf{p}+e\mathbf{A}')^2 &= (\mathbf{p}+e'\mathbf{A})^2+i\Sigma\sigma_3 \\
&\quad [(p_1+e'A_1)(p_2+e'A_2)-(p_2+e'A_2)(p_1+e'A_1)] \\
&= (\mathbf{p}+e'\mathbf{A})^2+he'(\boldsymbol{\sigma},\text{curl }\mathbf{A}).
\end{aligned}$$

Thus (15) becomes

$$\begin{aligned}
0 &= \Big[-(p_0+e'A_0)^2+(\mathbf{p}+e'\mathbf{A})^2+m^2c^2+e'h(\boldsymbol{\sigma},\text{curl }\mathbf{A}) \\
&\quad -ie'h\varrho_1\Big(\boldsymbol{\sigma},\text{grad }A_0+\frac{1}{c}\frac{\partial \mathbf{A}}{\partial t}\Big)\Big]\psi \\
&= [-(p_0+e'A_0)^2+(\mathbf{p}+e'\mathbf{A})^2+m^2c^2+e'h(\boldsymbol{\sigma},\mathbf{H})+ie'h\varrho_1(\boldsymbol{\sigma},\mathbf{E})]\psi,
\end{aligned}$$

where **E** and **H** are the electric and magnetic vectors of the field.

This differs from (1) by the two extra terms

$$\frac{eh}{c}(\boldsymbol{\sigma},\mathbf{H})+\frac{ieh}{c}\varrho_1(\boldsymbol{\sigma},\mathbf{E})$$

in F. These two terms, when divided by the factor $2m$, can be regarded as the additional potential energy of the electron due to its new degree of freedom. The electron will therefore behave as though it has a magnetic moment $eh/2mc \cdot \boldsymbol{\sigma}$ and an electric moment $ieh/2mc \cdot \varrho_1 \boldsymbol{\sigma}$. This magnetic moment is just that assumed in the spinning electron model. The electric moment, being a pure imaginary, we should not expect to appear in the model. It is doubtful whether the electric moment has any physical meaning, since the Hamiltonian in (14) that we started from is real, and the imaginary part only appeared when we multiplied it up in an artificial way in order to make it resemble the Hamiltonian of previous theories.

§ 5. The Angular Momentum Integrals for Motion in a Central Field

We shall consider in greater detail the motion of an electron in a central field of force. We put $\mathbf{A} = 0$ and $e'A_0 = V(r)$, an arbitrary function of the radius r, so that the Hamiltonian in (14) becomes

$$F \equiv p_0 + V + \varrho_1(\boldsymbol{\sigma}, \mathbf{p}) + \varrho_3 mc.$$

We shall determine the periodic solutions of the wave equation $F\psi = 0$, which means that p_0 is to be counted as a parameter instead of an operator; it is, in fact, just $1/c$ times the energy level.

We shall first find the angular momentum integrals of the motion. The orbital angular momentum $\mathbf{m}$ is defined by

$$\mathbf{m} = \mathbf{x} \times \mathbf{p},$$

and satisfies the following "Vertauschungs" relations

$$\left.\begin{array}{ll} m_1 x_1 - x_1 m_1 = 0, & m_1 x_2 - x_2 m_1 = ihx_3 \\ m_1 p_1 - p_1 m_1 = 0, & m_1 p_2 - p_2 m_1 = ihp_3 \\ \mathbf{m} \times \mathbf{m} = ih\mathbf{m}, & \mathbf{m}^2 m_1 - m_1 \mathbf{m}^2 = 0, \end{array}\right\}, \quad (17)$$

together with similar relations obtained by permuting the suffixes.

Also **m** commutes with r, and with p_r, the momentum canonically conjugate to r.

We have
$$m_1 F - F m_1 = \varrho_1\{m_1(\boldsymbol{\sigma}, \mathbf{p}) - (\boldsymbol{\sigma}, \mathbf{p})m_1\}$$
$$= \varrho_1(\boldsymbol{\sigma}, m_1\mathbf{p} - \mathbf{p}m_1)$$
$$= ih\varrho_1(\sigma_2 p_3 - \sigma_3 p_2),$$
and so
$$\mathbf{m}F - F\mathbf{m} = ih\varrho_1\, \boldsymbol{\sigma} \times \mathbf{p}. \tag{18}$$

Thus **m** is not a constant of the motion. We have further
$$\sigma_1 F - F\sigma_1 = \varrho_1\{\sigma_1(\boldsymbol{\sigma}, \mathbf{p}) - (\boldsymbol{\sigma}, \mathbf{p})\sigma_1\}$$
$$= \varrho_1(\sigma_1\boldsymbol{\sigma} - \boldsymbol{\sigma}\,\sigma_1, \mathbf{p})$$
$$= 2i\varrho_1(\sigma_3 p_2 - \sigma_2 p_3),$$

with the help of (8), and so
$$\boldsymbol{\sigma}F - F\boldsymbol{\sigma} = -2i\varrho_1\, \boldsymbol{\sigma} \times \mathbf{p}.$$
Hence
$$(\mathbf{m} + \tfrac{1}{2}h\boldsymbol{\sigma})F - F(\mathbf{m} + \tfrac{1}{2}h\boldsymbol{\sigma}) = 0.$$

Thus $\mathbf{m} + \tfrac{1}{2}h\boldsymbol{\sigma}$ ($= \mathbf{M}$ say) is a constant of the motion. We can interpret this result by saying that the electron has a spin angular momentum of $\tfrac{1}{2}h\boldsymbol{\sigma}$, which added to the orbital angular momentum **m**, gives the total angular momentum **M**, which is a constant of the motion.

The Vertauschungs relations (17) all hold when M's are written for the m's. In particular
$$\mathbf{M} \times \mathbf{M} = ih\mathbf{M} \quad \text{and} \quad \mathbf{M}^2 M_3 = M_3 \mathbf{M}^2.$$

M_3 will be an action variable of the system. Since the characteristic values of m_3 must be integral multiples of h in order that the wave function may be single-valued, the characteristic values of M_3 must be half odd integral multiples of h. If we put
$$\mathbf{M}^2 = (j^2 - \tfrac{1}{4})h^2, \tag{19}$$

j will be another quantum number, and the characteristic values of M_3 will extend from $(j-\frac{1}{2})h$ to $(-j+\frac{1}{2})h$*. Thus j takes integral values .

One easily verifies from (18) that $\mathbf{m}^2$ does not commute with F, and is thus not a constant of the motion. This makes a difference between the present theory and the previous spinning electron theory, in which $\mathbf{m}^2$ is constant, and defines the azimuthal quantum number k by a relation similar to (19). We shall find that our j plays the same part as the k of the previous theory.

§ 6. The Energy Levels for Motion in a Central Field

We shall now obtain the wave equation as a differential equation in r, with the variables that specify the orientation of the whole system removed. We can do this by the use only of elementary non-commutative algebra in the following way.

In formula (16) take $\mathbf{B} = \mathbf{C} = \mathbf{m}$. This gives

$$(\sigma, \mathbf{m})^2 = \mathbf{m}^2 + i(\sigma, \mathbf{m}\times\mathbf{m}) \qquad (20)$$
$$= (\mathbf{m}+\tfrac{1}{2}h\sigma)^2 - h(\sigma, \mathbf{m}) - \tfrac{1}{4}h^2\sigma^2 - h(\sigma, \mathbf{m})$$
$$= \mathbf{M}^2 - 2h(\sigma, \mathbf{m}) - \tfrac{3}{4}h^2.$$

Hence
$$\{(\sigma, \mathbf{m})+h\}^2 = \mathbf{M}^2 + \tfrac{1}{4}h^2 = j^2h^2.$$

Up to the present we have defined j only through j^2, so that we could now, if we liked, take jh equal to $(\sigma, \mathbf{m})+h$. This would not be convenient since we want j to be a constant of the motion while $(\sigma, \mathbf{m})+h$ is not, although its square is. We have, in fact, by another application of (16),

$$(\sigma, \mathbf{m})(\sigma, \mathbf{p}) = i(\sigma, \mathbf{m}\times\mathbf{p}),$$

hence $(\mathbf{m}, \mathbf{p}) = 0$, and similarly

$$(\sigma, \mathbf{p})(\sigma, \mathbf{m}) = i(\sigma, \mathbf{p}\times\mathbf{m}),$$

* See *Roy. Soc. Proc.*, A, vol. 111, p. 281 (1926).

252 SPECIAL RELATIVITY

so that

$$(\boldsymbol{\sigma}, \mathbf{m})(\boldsymbol{\sigma}, \mathbf{p}) + (\boldsymbol{\sigma}, \mathbf{p})(\boldsymbol{\sigma}, \mathbf{m}) = i\Sigma\sigma_1(m_2p_3 - m_3p_2 + p_2m_3 - p_3m_2)$$
$$= i\Sigma\sigma_1 \cdot 2ihp_1 = -2h(\boldsymbol{\sigma}, \mathbf{p}),$$

or

$$\{(\boldsymbol{\sigma}, \mathbf{m}) + h\}(\boldsymbol{\sigma}, \mathbf{p}) + (\boldsymbol{\sigma}, \mathbf{p})\{(\boldsymbol{\sigma}, \mathbf{m}) + h\} = 0.$$

Thus $(\boldsymbol{\sigma}, \mathbf{m}) + h$ anticommutes with one of the terms in F, namely, $\varrho_1(\boldsymbol{\sigma}, \mathbf{p})$, and commutes with the other three. Hence $\varrho_3\{(\boldsymbol{\sigma}, \mathbf{m}) + h\}$ commutes with all four, and is therefore a constant of the motion. But the square of $\varrho_3\{(\boldsymbol{\sigma}, \mathbf{m}) + h\}$ must also equal j^2h^2. We therefore take

$$jh = \varrho_3\{(\boldsymbol{\sigma}, \mathbf{m}) + h\}. \tag{21}$$

We have, by a further application of (16),

$$(\boldsymbol{\sigma}, \mathbf{x})(\boldsymbol{\sigma}, \mathbf{p}) = (\mathbf{x}, \mathbf{p}) + i(\boldsymbol{\sigma}, \mathbf{m}).$$

Now a permissible definition of p_r is

$$(\mathbf{x}, \mathbf{p}) = rp_r + ih,$$

and from (21)

$$(\boldsymbol{\sigma}, \mathbf{m}) = \varrho_3 jh - h.$$

Hence

$$(\boldsymbol{\sigma}, \mathbf{x})(\boldsymbol{\sigma}, \mathbf{p}) = rp_r + i\varrho_3 jh. \tag{22}$$

Introduce the quantity ε defined by

$$r\varepsilon = \varrho_1(\boldsymbol{\sigma}, \mathbf{x}). \tag{23}$$

Since r commutes with ϱ_1 and with $(\boldsymbol{\sigma}, \mathbf{x})$, it must commute with ε. We thus have

$$r^2\varepsilon^2 = [\varrho_1(\boldsymbol{\sigma}, \mathbf{x})]^2 = (\boldsymbol{\sigma}, \mathbf{x})^2 = \mathbf{x}^2 = r^2$$

or

$$\varepsilon^2 = 1.$$

Since there is symmetry between $\mathbf{x}$ and $\mathbf{p}$ so far as angular momentum is concerned, $\varrho_1(\boldsymbol{\sigma}, \mathbf{x})$, like $\varrho_1(\boldsymbol{\sigma}, \mathbf{p})$, must commute with

DIRAC: QUANTUM THEORY OF THE ELECTRON 253

M and j. Hence ε commutes with **M** and j. Further, ε must commute with p_r, since we have

$$(\boldsymbol{\sigma}, \mathbf{x})(\mathbf{x}, \mathbf{p}) - (\mathbf{x}, \mathbf{p})(\boldsymbol{\sigma}, \mathbf{x}) = ih(\boldsymbol{\sigma}, \mathbf{x}),$$

which gives

$$r\varepsilon(rp_r + ih) - (rp_r + ih)r\varepsilon = ihr\varepsilon,$$

which reduces to

$$\varepsilon p_r - p_r \varepsilon = 0.$$

From (22) and (23) we now have

$$r\varepsilon\varrho_1(\boldsymbol{\sigma}, \mathbf{p}) = rp_r + i\varrho_3 jh$$

or

$$\varrho_1(\boldsymbol{\sigma}, \mathbf{p}) = \varepsilon p_r + i\varepsilon p_3 jh/r.$$

Thus

$$F = p_0 + V + \varepsilon p_r + i\varepsilon\varrho_3 jh/r + \varrho_3 mc. \tag{24}$$

Equation (23) shows that ε anticommutes with ϱ_3. We can therefore by a canonical transformation (involving perhaps the x's and p's as well as the σ's and ϱ's) bring ε into the form of the ϱ_2 of § 2 without changing ϱ_3, and without changing any of the other variables occurring on the right-hand side of (24), since these other variables all commute with ε. $i\varepsilon\varrho_3$ will now be of the form $i\varrho_2\varrho_3 = -\varrho_1$, so that the wave equation takes the form

$$F\psi \equiv [p_0 + V + \varrho_2 p_r - \varrho_1 jh/r + \varrho_3 mc]\psi = 0.$$

If we write this equation out in full, calling the components of ψ referring to the first and third rows (or columns) of the matrices ψ_α and ψ_β respectively, we get

$$(F\psi)_\alpha \equiv (p_0 + V)\psi_\alpha - h\frac{\partial}{\partial r}\psi_\beta - \frac{jh}{r}\psi_\beta + mc\psi_\alpha = 0,$$

$$(F\psi)_\beta \equiv (p_0 + V)\psi_\beta + h\frac{\partial}{\partial r}\psi_\alpha - \frac{jh}{r}\psi_\alpha - mc\psi_\beta = 0.$$

254 SPECIAL RELATIVITY

The second and fourth components give just a repetition of these two equations. We shall now eliminate ψ_α. If we write hB for $p_0 + V + mc$, the first equation becomes

$$\left(\frac{\partial}{\partial r} + \frac{j}{r}\right)\psi_\beta = B\psi_\alpha,$$

which gives on differentiating

$$\frac{\partial^2}{\partial r^2}\psi_\beta + \frac{j}{r}\frac{\partial}{\partial r}\psi_\beta - \frac{j}{r^2}\psi_\beta = B\frac{\partial}{\partial r}\psi_\alpha + \frac{\partial B}{\partial r}\psi_\alpha$$

$$= \frac{B}{h}\left[-(p_0 + V - mc)\psi_\alpha + \frac{jh}{r}\psi_\alpha\right] + \frac{1}{h}\frac{\partial V}{\partial r}\psi_\alpha$$

$$= -\frac{(p_0+V)^2 - m^2c^2}{h^2}\psi_\beta + \left(\frac{j}{r} + \frac{1}{Bh}\frac{\partial V}{\partial r}\right)\left(\frac{\partial}{\partial r} + \frac{j}{r}\right)\psi_\beta.$$

This reduces to

$$\frac{\partial^2}{\partial r^2}\psi_\beta + \left[\frac{(p_0+V)^2 - m^2c^2}{h^2} - \frac{j(j+1)}{r^2}\right]\psi_\beta - \frac{1}{Bh}\frac{\partial V}{\partial r}\left(\frac{\partial}{\partial r} + \frac{j}{r}\right)\psi_\beta = 0. \tag{25}$$

The values of the parameter p_0 for which this equation has a solution finite at $r = 0$ and $r = \infty$ are $1/c$ times the energy levels of the system. To compare this equation with those of previous theories, we put $\psi_\beta = r\chi$, so that

$$\frac{\partial^2}{\partial r^2}\chi + \frac{2}{r}\frac{\partial}{\partial r}\chi + \left[\frac{(p_0+V)^2 - m^2c^2}{h^2} - \frac{j(j+1)}{r^2}\right]\chi$$
$$- \frac{1}{Bh}\frac{\partial V}{\partial r}\left(\frac{\partial}{\partial r} + \frac{j+1}{r}\right)\chi = 0. \tag{26}$$

If one neglects the last term, which is small on account of B being large, this equation becomes the same as the ordinary Schroedinger equation for the system, with relativity correction included. Since j has, from its definition, both positive and negative integral characteristic values, our equation will give twice as many energy levels when the last term is not neglected.

We shall now compare the last term of (26), which is of the same order of magnitude as the relativity correction, with the spin correction given by Darwin and Pauli. To do this we must eliminate the $\partial \chi/\partial r$ term by a further transformation of the wave function. We put

$$\chi = B^{-1/2}\chi_1,$$

which gives

$$\frac{\partial^2}{\partial r^2}\chi_1 + \frac{2}{r}\frac{\partial}{\partial r}\chi_1 + \left[\frac{(p_0+V)^2-m^2c^2}{h^2} - \frac{j(j+1)}{r^2}\right]\chi_1$$
$$+ \left[\frac{1}{Bh}\frac{j}{r}\frac{\partial V}{\partial r} - \frac{1}{2}\frac{1}{Bh}\frac{\partial^2 V}{\partial r^2} + \frac{1}{4}\frac{1}{B^2h^2}\left(\frac{\partial V}{\partial r}\right)^2\right]\chi_1 = 0. \quad (27)$$

The correction is now, to the first order of accuracy,

$$\frac{1}{Bh}\left(\frac{j}{r}\frac{\partial V}{\partial r} - \frac{1}{2}\frac{\partial^2 V}{\partial r^2}\right),$$

where $Bh = 2mc$ (provided p_0 is positive). For the hydrogen atom we must put $V = e^2/cr$. The first order correction now becomes

$$-\frac{e^2}{2mc^2r^3}(j+1). \quad (28)$$

If we write $-j$ for $j+1$ in (27), we do not alter the terms representing the unperturbed system, so

$$\frac{e^2}{2mc^2r^3}j \quad (28')$$

will give a second possible correction for the same unperturbed term.

In the theory of Pauli and Darwin, the corresponding correcting term is

$$\frac{e^2}{2mhc^2r^3}(\boldsymbol{\sigma}, \mathbf{m})$$

when the Thomas factor $\frac{1}{2}$ is included. We must remember that in the Pauli–Darwin theory, the resultant orbital angular momentum k plays the part of our j. We must define k by

$$\mathbf{m}^2 = k(k+1)h^2$$

instead of by the exact analogue of (19), in order that it may have integral characteristic values, like j. We have from (20)

$$(\boldsymbol{\sigma}, \mathbf{m})^2 = k(k+1)h^2 - h(\boldsymbol{\sigma}, \mathbf{m})$$

or

$$\{(\boldsymbol{\sigma}, \mathbf{m}) + \tfrac{1}{2}h\}^2 = (k+\tfrac{1}{2})^2 h^2,$$

hence

$$(\boldsymbol{\sigma}, \mathbf{m}) = kh \quad \text{or} \quad -(k+1)h.$$

The correction thus becomes

$$\frac{e^2}{2mc^2r^3}k \quad \text{or} \quad -\frac{e^2}{2mc^2r^3}(k+1),$$

which agrees with (28) and (28'). The present theory will thus, in the first approximation, lead to the same energy levels as those obtained by Darwin, which are in agreement with experiment.

NOTES ON EXTRACT 9

THIS paper gives a good idea of the difficulties of observation in the borderland of quantum mechanics and relativity, especially in the thirties. Anderson was working in ignorance of Dirac's theoretical prediction of the positron and a good deal of his paper is taken up with assessment of positron tracks on the false assumption that they are protons. The difficulties into which this runs become clear in the section headed "Associated tracks"; especially the large mass of the proton is mentioned.

EXTRACT 9[*]

Energies of Cosmic-ray Particles

By CARL D. ANDERSON

California Institute of Technology
(Received June 28, 1932)

> Cloud chamber photographs of cosmic-ray tracks in a magnetic field
> up to 17,000 gauss are shown. On the assumption that the particles
> producing the tracks are traveling downward through the chamber
> rather than upward, particles of positive charge appear as well as
> electrons. From the specific ionization along the track it is concluded
> that the positives are protons, and are not nuclei of charge greater
> than unity. No evidence is uncovered demanding the introduction of
> a neutron for cosmic-ray phenomena. Eight examples of associated
> tracks are shown. Energies range from below 10^6 electron-volts to
> values in a few cases of the order of 10^9 electron-volts. Energy values
> for 70 tracks are listed. The scattering of cosmic particles in traversing
> a 6.0 mm lead plate is measured.

AN AUTOMATIC, vertical Wilson expansion chamber operating in a strong magnetic field up to 21,000 gauss was designed for a study of the high-energy corpuscles associated with cosmic rays. The expansion chamber itself is 15 cm in diameter and has a depth of 2 cm. The axis of the piston lies in a horizontal plane in order to effect the most favorable position for photographing the cosmic-ray tracks. The magnetic field was found by direct measurement to be homogeneous to within 10 percent throughout the volume of the chamber. Photographs are taken through a hole in the pole piece of the magnet along the lines of force, thus revealing a particle deflected by the magnetic field as an arc of a circle on the pho-

[* *Phys. Rev.* **41** (2), 405 (1932).]

tographic film. A description of the apparatus will follow in a later publication.

A brief discussion of the results has been published jointly with Professor R. A. Millikan,[1] to whom the writer is indebted for suggesting the investigation and cooperating in planning it, and whose keen interest throughout the course of the work has been of the greatest value.

Of the 3000 photographs taken to date, 62 show cosmic-ray tracks of length sufficient for energy measurements, 19 of which are here reproduced. None of the photographs are reproduced in this extract with addition of 3 photographs taken for test purposes. The photographs are 7/10 full size. The dark back ground in several of the photographs is due to light scattered by the back wall of the expansion chamber.

In the test photograph of Fig. 2 appear tracks in air of the secondary electrons ejected by gamma-rays from Ra. Figs. 3 and 4 show alpha-particle trajectories.

The remainder of the tracks shown, Figs. 5–23, are ascribed to cosmic rays because of their high energy. Cosmic-ray tracks are in all cases readily distinguishable from alpha-particle tracks due to the very much greater specific ionization of the latter.

Positively Charged Particles

A charged particle will be deviated by the magnetic field into an arc of a circle. The sense of rotation in the chamber as viewed in the photographs will be clockwise for a particle of negative charge, and counter-clockwise for one of positive charge. The sign of the charge can be ascertained only if the direction of motion of the particle is known. It is assumed here that the particles are traveling downward through the chamber. The small degree of scattering to be expected for high-energy particles, combined with the known fact that the rays come in from above, appears to justify

[1] Millikan and Anderson, *Phys. Rev.* 40, 325 (1932).

this assumption. In Fig. 1 it is seen that the cosmic-ray tracks prefer the vertical direction over the horizontal indicating that backward or large angle scattering is infrequent.

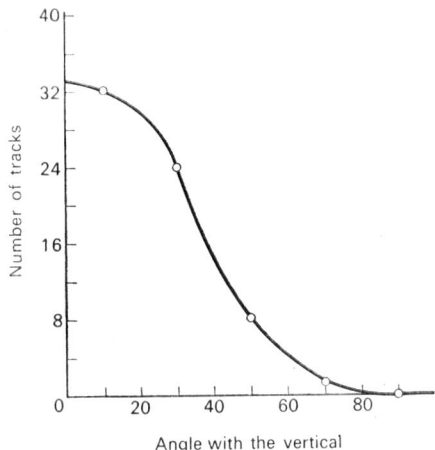

Angle with the vertical

FIG. 1. The number of tracks per unit solid angle as a function of the angle with the vertical. Only those tracks are included whose curvature in the magnetic field is sufficiently small to allow determination of the direction to be made. Therefore no electrons of energy less than 100×10^6 volts are included. This space-distribution in showing a large percentage of nearly vertical tracks differs from that reported by Skolbeltzyn.[2]

In many instances on the above assumption as to the direction of motion, the tracks are deviated in a sense to indicate the presence of positively charged particles as well as electrons.

Specific Ionization

A few tracks photographed in an atmosphere of helium with 5 percent air show about 14 ion pairs per cm at standard pressure, a value close to that found for electrons of about 10^6 volts energy

[2] Skobeltzyn, *C.R.* **194,** 118 (1932).

from Ra gamma-rays, which show about 13 ion pairs per cm in the same gas. The specific ionization for an electron remains practically independent of its energy for energies ranging from about 300,000 volts to several hundred million volts. For a given velocity, protons and electrons ionize the same, and for high energies where the velocities of both the electrons and the protons are of the order of the velocity of light it becomes impossible to distinguish electrons from protons by their ionization. Only at lower energies where the proton velocities are appreciably less than the velocity of light will protons show an appreciably greater specific ionization than electrons of the same energy. Nuclei of higher atomic charge would, for a given velocity, produce many more ions per cm, the specific ionization being to a first approximation proportional to the square of the charge on the nucleus.

The specific ionization along the tracks showing positives is in most instances not much greater than that for the electrons. It is concluded, therefor; that the positives can only be protons, and cannot themselves represent nuclei of higher atomic number than unity.

The projection of whole nuclei by the penetrating radiation produced in the bombardment of beryllium with alpha-particles has been reported in recent experiments. For the explanation of this fact, on the basis of the conservation laws, Chadwick[3] postulates a neutron. For the interpretation of the cosmic-ray effects so far observed such a neutron is not demanded on the basis of the energy–momentum arguments which apply in the experiments of Chadwick. Further work will show if the associated tracks of cosmic rays represent an effect similar, but on a higher energy scale to the disintegration tracks photographed by Feather[4] in the neutron experiments, though the frequent occurrence of electron tracks in the cosmic-ray experiments seems to indicate a different type of phenomenon.

[3] Chadwick, *Nature* **129,** 312 (1932).
[4] Feather, *Proc. Roy. Soc. A* **136,** 709 (1932).

A possibility to be borne in mind is that, in rare cases, the tracks of curvature that indicate positives might be in reality electrons scattered backwards by the material underneath the chamber and are traversing it from bottom to top. Precise data on the specific ionization of the low-energy positives will distinguish, however, between downward positives and upward negatives.

Associated Tracks

A well-known characteristic of the cosmic tracks is their tendency to occur in groups.[5, 6, 7] Of the 55 photographs showing cosmic tracks, 7 show double tracks and 1 shows three tracks.

In general, for paired tracks, the energy of one of the associated pair is considerably less than that of the other, in some instances 10^6 volts and less. One of the associated pair is also in all cases definitely an electron.

The associated tracks have been assumed to be due to the simultaneous ejection by a photon of two particles from an atomic nucleus.[1]

Another effect which may give rise to associated tracks is a close encounter between a cosmic particle and an electron. Fig. 20 is an example of an encounter of this type, the encounter taking place in the wall. For such an encounter where an electron of high energy (energy $\gg mc^2$) produces a secondary track, giving to the secondary electron an energy E, the angle θ between the primary and secondary electron is given by $\tan \theta = (2mc^2/E)^{1/2}$. The two tracks of Fig. 10 might represent an effect of this type, the relation above being satisfied within experimental uncertainty. The possibility also exists that a proton may by direct impact give to an electron energy sufficient to produce a secondary track. It is pointed out, however, that on the basis of the conservation laws, due to the

[5] Skobelzyn, *Zeits. f. Physik* **54,** 686 (1929).
[6] Auger and Skobelzyn, *C.R.* **189,** 55 (1929).
[7] Locher, *Phys. Rev.* **39,** 883 (1932).

relatively large proper mass of a proton, it is difficult to explain the associated tracks as a binary collision between a proton and an electron because of the prohibitively high energy which would have to be assigned to the proton in this case to account for the electron energies observed. To produce an electron of 100×10^6 volts would require a proton energy of 10^{10} volts. In both Figs. 8 and 11 one of the associated pair has a curvature to indicate a positive, and therefore, if these curvatures are correctly measured these cases cannot represent such encounters. Even if these are incorrectly measured and are in reality electrons then for Fig. 11 the value of θ calculated for an energy of 27×10^6 volts is $11°$, which would agree with the angle indicated by the photograph within experimental uncertainty. But for the case of Fig. 8, the calculated value of $17°$ and the measured value of $25°$ are in conflict. Again the associated pair of Fig. 5 can not possibly be interpreted as due to a binary collision. The hypothesis of simple binary collisions is inadequate to account for all the associated tracks. Furthermore, encounters of this type in which a large amount of energy is transferred to an electron are not to be expected frequently, and the abundance of associated tracks coupled with the fact that positives as well as electrons appear, is strong evidence that the associated tracks represent a quite different phenomenon, i.e., the ejection of two particles from a nucleus.

Scattering of the Cosmic Particles

Certain of the tracks, Figs. 11 and 17, show sudden though very small deflections identical in appearance with the deflections observed in alpha-particle tracks due to nuclear encounters, but which are to be expected in the gas only rarely on the basis of the present scattering laws for high-energy electrons or protons. The deflections in some instances represent scattering from the walls of the chamber. An example of large angle scattering from a lead surface is shown in Fig. 8. In Fig. 10 another instance of large angle scattering is shown, the electron being scattered by the glass wall of the chamber.

Experiments are now in progress to study the scattering of cosmic particles in lead. Figs. 22 and 23 show particles traversing 6·0 mm of lead, the angle of scattering being in each case readily measurable.

Energies

The energies of the cosmic particles as determined from the radii of curvature of the tracks range from values below 10^6 electron-volts to, in a few cases, values of the order of 10^9 electron-volts. The greater part of them, however, have energies below 500×10^6 volts.

Precautions were taken to reduce to a minimum the effects of air movements in the expansion chamber which tend to distort the tracks. The energies of the higher energy tracks, i.e., protons of energies of the order of 500×10^6 volts and electrons of the order of 10^9 volts, can be determined only roughly due to the small curvature in the magnetic field.

Table I gives the distribution in energy of the electrons and protons. The number of protons and the number of electrons in various energy ranges are listed. It is to be noted that there may be many more electrons in the energy range below 10^6 volts than those listed. Since this includes the energy region of radio-activity, it is impossible to distinguish between electrons from radio-active sources and low-energy electrons due to cosmic rays. Therefore only those very low-energy electrons which are associated with other cosmic-ray tracks, and are definitely to be attributed to cosmic rays are listed in Table I.

There are in addition 5 tracks whose sign of charge is doubtful due to the lack of an appreciable curvature in the magnetic field. For these it is possible only to assign a lower limit to the energy, i.e., 450×10^6 volts on the supposition that they are electrons and

SPECIAL RELATIVITY

TABLE I. *Energy distribution*

Number of electrons	Number of protons	Energy range in electron-volts
2	0	Below 10^6
4	0	From 10^6 to 10×10^6
6	1	From 10×10^6 to 20×10^6
4	1	From 20×10^6 to 30×10^6
3	3	From 30×10^6 to 50×10^6
1	3	From 50×10^6 to 100×10^6
5	5	From 100×10^6 to 200×10^6
4	2	From 200×10^6 to 300×10^6
5	2	From 300×10^6 to 400×10^6
1	4	From 400×10^6 to 500×10^6
3	3 {	From 500×10^6 to 700×10^6
2		From 700×10^6 to 1000×10^6

TABLE II. *Energies of the associated tracks in electron-volts*

Group 1	Proton Electron	130×10^6 120×10^6	See Fig. 5
Group 2	Electron Probably proton Probably proton	30×10^6 150×10^6 400×10^6	See Fig. 6
Group 3	Probably proton Electron	450×10^5 27×10^6	See Fig. 11
Group 4	Proton Electron	20×10^6 4×10^5	See Fig. 7
Group 5	Probably proton Electron	$\sim 100 \times 10^6$ 11×10^6	See Fig. 8
Group 6	Electron Electron	170×10^6 11×10^6	See Fig. 10
Group 7	Electron Electron	180×10^6 2×10^6	See Fig. 9
Group 8	Electron Electron or proton	4×10^5 $\sim 400 \times 10^6$	See Fig. 20

100×10^6 volts on the supposition that they are protons. There is one track, Fig. 21 to which an energy in excess of 10^9 volts must be assigned whether it is assumed a proton or electron.

In Table II are given the energies of the 8 groups of associated tracks.

I wish to thank Mr. Seth H. Neddermeyer for assistance in the measurement of the photographs.

NOTES ON EXTRACT 10

ABOUT the first one-third of this exceptionally important paper is printed here. The later parts deal respectively with the reduction of representations (up to a factor) to two-valued representations, and the reduction of the representations to those of a "little group". These are of a rather more technical nature.

EXTRACT 10[*]

On Unitary Representations of the Inhomogeneous Lorentz Group*

By E. WIGNER

(Received December 22, 1937)

1. Origin and Characterization of the Problem

It is perhaps the most fundamental principle of Quantum Mechanics that the system of states forms a *linear manifold*,[1] in which a unitary *scalar product* is defined.[2] The states are generally represented by wave functions[3] in such a way that ϕ and constant mul-

[* *Ann. of Math.* **40**, 149 (1939).]

* Parts of the present paper were presented at the Pittsburgh Symposium on Group Theory and Quantum Mechanics. Cf. *Bull. Amer. Math. Soc.*, **41**, p. 306, 1935.

[1] The possibility of a future non linear character of the quantum mechanics must be admitted, of course. An indication in this direction is given by the theory of the positron, as developed by P. A. M. Dirac *(Proc. Camb. Phil. Soc.* **30**, 150, 1934, cf. also W. Heisenberg, *Zeits. f. Phys.* **90**, 209, 1934; **92**, 623, 1934; W. Heisenberg and H. Euler, *ibid.* **98**, 714, 1936, and R. Serber, *Phys. Rev.* **48**, 49, 1935; **49**, 545, 1936) which does not use wave functions and is a non linear theory.

[2] Cf. P. A. M. Dirac, *The Principles of Quantum Mechanics*, Oxford, 1935, Chapters I and II; J. v. Neumann, *Mathematische Grundlagen der Quantenmechanik*, Berlin, 1932, pages 19–24.

[3] The wave functions represent throughout this paper states in the sense of the "Heisenberg picture," i. e. a single wave function represents the state for all past and future. On the other hand, the operator which refers to a measurement at a certain time t contains this t as a parameter. (Cf., e.g., Dirac, loc. cit. ref. 2, pages 115–123). One obtains the wave function $\phi_s(t)$ of the Schrödinger picture from the wave function ϕ_H of the Heisenberg picture by $\phi_s(t) = \exp(-iHt/h)\phi_H$. The operator of the Heisenberg picture is $Q(t) =$

tiples of ϕ represent the same physical state. It is possible, therefore to normalize the wave function, i.e., to multiply it by a constant factor such that its scalar product with itself becomes 1. Then, only a constant factor of modulus 1, the so-called phase, will be left undetermined in the wave function. The linear character of the wave function is called the superposition principle. The square of the modulus of the unitary scalar product (ψ, ϕ) of two normalized wave functions ψ and ϕ is called the transition probability from the state ψ into ϕ, or conversely. This is supposed to give the probability that an experiment performed on a system in the state ϕ, to see whether or not the state is ψ, gives the result that it is ψ. If there are two or more different experiments to decide this (e.g., essentially the same experiment, performed at different times) they are all supposed to give the same result, i.e., the transition probability has an invariant physical sense.

The wave functions form a description of the physical state, not an invariant however, since the same state will be described in different coördinate systems by different wave functions. In order to put this into evidence, we shall affix an index to our wave functions, denoting the Lorentz frame of reference for which the wave function is given. Thus ϕ_l and $\phi_{l'}$, represent the same state, but they are different functions. The first is the wave function of the state in the coordinate system l, the second in the coördinate system l'. If $\phi_l = \psi_{l'}$ the state ϕ behaves in the coördinate system l exactly as ψ behaves in the coördinate system l'. If ϕ_l is given, all $\phi_{l'}$ are determined up to a constant factor. Because of the invariance of the transition probability we have

$$|(\phi_l, \psi_l)|^2 = |(\phi_{l'}, \psi_{l'})|^2 \tag{1}$$

exp $(iHt/h)\, Q \exp(-iHt/h)$, where Q is the operator in the Schrödinger picture which does not depend on time. Cf. also E. Schrödinger, *Sitz. d. Kön. Preuss. Akad.*, p. 418, 1930.

The wave functions are complex quantities and the undetermined factors in them are complex also. Recently attempts have been made toward a theory with real wave functions. Cf. E. Majorana, *Nuovo Cim.* **14**, 171, 1937 and P. A. M. Dirac, *Proc. Camb. Phil. Soc.* **35**, 416, 1939.

and it can be shown[4] that the aforementioned constants in the $\phi_{l'}$ can be chosen in such a way that the $\phi_{l'}$ are obtained from the ϕ_l by a linear unitary operation, depending, of course, on l and l'

$$\varphi_{l'} = D(l', l)\phi_l. \tag{2}$$

The unitary operators D are determined by the physical content of the theory up to a constant factor again, which can depend on l and l'. Apart from this constant however, the operations $D(l', l)$ and $D(l'_1, l_1)$ must be identical if l' arises from l by the same Lorentz transformation, by which l'_1 arises from l_1. If this were not true, there would be a real difference between the frames of reference l and l_1. Thus the unitary operator $D(l', l) = D(L)$ is in every Lorentz invariant quantum mechanical theory (apart from the constant factor which has no physical significance) completely determined by the Lorentz transformation L which carries l into $l' = Ll$. One can write, instead of (2)

$$\phi_{Ll} = D(L)\phi_l. \tag{2a}$$

By going over from a first system of reference l to a second $l' = L_1 l$ and then to a third $l'' = L_2 L_1 l$ or directly to the third $l'' = (L_2 L_1)l$, one must obtain—apart from the above mentioned constant—the same set of wave functions. Hence from

$$\phi_{l''} = D(l'', l') \, D(l', l)\phi_l$$
$$\phi_{l''} = D(l'', l)\phi_l$$

it follows

$$D(l'', l') \, D(l', l) = \omega D(l'', l) \tag{3}$$

or

$$D(L_2)D(L_1) = \omega D(L_2 L_1), \tag{3a}$$

where ω is a number of modulus 1 and can depend on L_2 and L_1. Thus the $D(L)$ form, up to a factor, a representation of the inhomogeneous Lorentz group by linear, unitary operators.

[4] E. Wigner, *Gruppentheorie und ihre Anwendungen auf die Quantenmechanik der Atomspektren*, Braunschweig, 1931, pages 251–254.

We see thus[5] that there corresponds to every invariant quantum mechanical system of equations such a representation of the inhomogeneous Lorentz group. This representation, on the other hand, though not sufficient to replace the quantum mechanical equations entirely, can replace them to a large extent. If we knew, e.g., the operator K corresponding to the measurement of a physical quantity at the time $t = 0$, we could follow up the change of this quantity throughout time. In order to obtain its value for the time $t = t_1$, we could transform the original wave function ϕ_2 by $D(l', l)$ to a coördinate system l' the time scale of which begins a time t_1 later. The measurement of the quantity in question in this coördinate system for the time 0 is given—as in the original one—by the operator K. This measurement is indentical, however, with the measurement of the quantity at time t_1 in the original system. One can say that the representation can replace the equation of motion, it cannot replace, however, connections holding between operators at one instant of time.

It may be mentioned, finally, that these developments apply not only in quantum mechanics, but also to all linear theories, e.g., the Maxwell equations in empty space. The only difference is that there is no arbitrary factor in the description and the ω can be omitted in (3a) and one is led to real representations instead of representations up to a factor. On the other hand, the unitary character of the representation is not a consequence of the basic assumptions.

The increase in generality, obtained by the present calculus, as compared with the usual tensor theory, consists in that no assumptions regarding the field nature of the underlying equations are necessary. Thus more general equations, as far as they exist (e.g., in which the coördinate is quantized, etc.) are also included in the present treatment. It must be realized, however, that some assumptions concerning the continuity of space have been made by assuming Lorentz frames of reference in the classical sense. We

[5] E. Wigner, loc. cit., Chapter XX.

should like to mention, on the other hand, that the previous remarks concerning the time-parameter in the observables, have only an explanatory character, and we do not make assumptions of the kind that measurements can be performed instantaneously.

We shall endeavor, in the ensuing sections, to determine all the continuous[6] unitary representations up to a factor of the inhomogeneous Lorentz group, i.e., all continuous systems of linear, unitary operators satisfying (3a).

2. Comparison With Previous Treatments and Some Immediate Simplifications

A. Previous Treatments

The representations of the Lorentz group have been investigated repeatedly. The first investigation is due to Majorana,[7] who in fact found all representations of the class to be dealt with in the present work excepting two sets of representations. Dirac[8] and Proca[8] gave more elegant derivations of Majorana's results and brought them into a form which can be handled more easily. Klein's work[9] does not endeavor to derive irreducible representations and seems to be in a less close connection with the present work.

The difference between the present paper and that of Majorana and Dirac lies—apart from the finding of new representations—mainly in its greater mathematical rigor. Majorana and Dirac freely use the notion of infinitesimal operators and a set of func-

---

[6] The exact definition of the continuous character of a representation up to a factor will be given in Section 5A. The definition of the inhomogeneous Lorentz group is contained in Section 4A.

[7] E. Majorana, *Nuovo Cim.* **9**, 335, 1932.

[8] P. A. M. Dirac, *Proc. Roy. Soc.* A, **155**, 447, 1936; Al. Proca, *J. de Phys. Rad.* **7**, 347, 1936.

[9] Klein, *Arkiv f. Matem. Astr. och Fysik*, **25**A, No. 15, 1936. I am indebted to Mr. Darling for an interesting conversation on this paper.

tions to all members of which every infinitesimal operator can be applied. This procedure cannot be mathematically justified at present, and no such assumption will be used in the present paper. Also the conditions of reducibility and irreducibility could be, in general, somewhat more complicated than assumed by Majorana and Dirac. Finally, the previous treatments assume from the outset that the space and time coördinates will be continuous variables of the wave function in the usual way. This will not be done, of course, in the present work.

B. Some Immediate Simplifications

Two representations are *physically equivalent* if there is a one to one correspondence between the states of both which is 1. invariant under Lorentz transformations and 2. of such a character that the transition probabilities between corresponding states are the same.

It follows from the second condition[5] that there either exists a unitary operator S by which the wave functions $\Phi^{(2)}$ of the second representation can be obtained from the corresponding wave functions $\Phi^{(1)}$ of the first representation

$$\Phi^{(2)} = S\Phi^{(1)} \tag{4}$$

or that this is true for the conjugate imaginary of $\Phi^{(2)}$. Although, in the latter case, the two representations are still equivalent physically, we shall, in keeping with the mathematical convention, not call them equivalent.

The first condition now means that if the states $\Phi^{(1)}, \Phi^{(2)} = S\Phi^{(1)}$ correspond to each other in one coördinate system, the states $D^{(1)}(L)\Phi^{(1)}$ and $D^{(2)}(L)\Phi^{(2)}$ correspond to each other also. We have then

$$D^{(2)}(L)\Phi^{(2)} = SD^{(1)}(L)\Phi^{(1)} = SD^{(1)}(L)S^{-1}\Phi^{(2)}. \tag{4a}$$

As this shall hold for every $\Phi^{(2)}$, the existence of a unitary S which transforms $D^{(1)}$ into $D^{(2)}$ is the condition for the equivalence of these two representations. Equivalent representations are not considered to be really different and it will be sufficient to find one sample from every infinite class of equivalent representations.

If there is a closed linear manifold of states which is invariant under all Lorentz transformations, i.e. which contains $D(L)\psi$ if it contains ψ, the linear manifold perpendicular to this one will be invariant also. In fact, if ϕ belongs to the second manifold, $D(L)\phi$ will be, on account of the unitary character of $D(L)$, perpendicular to $D(L)\psi'$ if ψ' belongs to the first manifold. However, $D(L^{-1})\psi$ belongs to the first manifold if ψ does and thus $D(L)\phi$ will be orthogonal to $D(L) D(L^{-1})\psi = \omega\psi$ i.e. to all members of the first manifold and belong itself to the second manifold also. The original representation then "decomposes" into two representations, corresponding to the two linear manifolds. It is clear that, conversely, one can form a representation, by simply "adding" several other representations together, i.e. by considering as states linear combinations of the states of several representations and assume that the states which originate from different representations are perpendicular to each other.

Representations which are equivalent to sums of already known representations are not really new and, in order to master all representations, it will be sufficient to determine those, out of which all others can be obtained by "adding" a finite or infinite number of them together.

Two simple theorems shall be mentioned here which will be proved later (Sections 7A and 8C respectively). The first one refers to unitary representations of any closed group, the second to irreducible unitary representations of any (closed or open) group.

The representations of a closed group by unitary *operators* can be transformed into the sum of unitary representations with matrices of finite dimensions.

Given two non equivalent irreducible unitary representations of an arbitrary group. If the scalar product between the wave functions is invariant under the operations of the group, the wave functions belonging to the first respresentation are orthogonal to all wave functions belonging to the second representation.

C. *Classification of unitary representations according to von Neumann and Murray*[10]

Given the operators $D(L)$ of a unitary representation, or a representation up to a factor, one can consider the algebra of these operators, i.e. all linear combinations

$$a_1D(L_1)+a_2D(L_2)+a_3D(L_3)+ \ldots$$

of the $D(L)$ and all limits of such linear combinations which are bounded operators. According to the properties of this representation algebra, three classes of unitary representations can be distinguished.

The first class of *irreducible* representations has a representation algebra which contains all bounded operators, i.e. if ψ and ϕ are two arbitrary states, there is an operator A of the representation algebra for which $A\psi = \phi$ and $A\psi' = 0$ if ψ' is orthogonal to ψ. It is clear that the center of the algebra contains only the unit operator and multiples thereof. In fact, if C is in the center one can decompose $C\psi = \alpha\psi+\psi'$ so that ψ' shall be orthogonal to ψ. However, ψ' must vanish since otherwise C would not commute with the operator which leaves ψ invariant and transforms every function orthogonal to it into 0. For similar reasons, α must be the same for all ψ. For irreducible representations there is no closed linear manifold of states, (excepting the manifold of all states) which is invariant under all Lorentz transformations. In fact, according to the above definition, a ϕ' arbitrarily close to any ϕ can be represented by a finite linear combination

$$a_1D(L_1)\psi+a_2D(L_2)\psi+ \ldots +a_nD(L_n)\psi.$$

[10] F. J. Murray and J. v. Neumann, *Ann. of Math.* **37**, 116, 1936; J. v. Neumann, to be published soon.

Hence, a closed linear invariant manifold contains every state if it contains one. This is, in fact, the more customary definition for irreducible representations and the one which will be used subsequently. It is well known that all finite dimensional representations are sums of irreducible representations. This is not true,[10] in general, in an infinite number of dimensions.

The second class of representations will be called *factorial*. For these, the center of the representation algebra still contains only multiples of the unit operator. Clearly, the irreducible representations are all factorial, but not conversely. For finite dimensions, the factorial representations may contain one irreducible representation several times. This is also possible in an infinite number of dimensions, but in addition to this, there are the "continuous" representations of Murray and von Neumann.[10] These are not irreducible as there are invariant linear manifolds of states. On the other hand, it is impossible to carry the decomposition so far as to obtain as parts only irreducible representations. In all the examples known so far, the representations into which these continuous representations can be decomposed, are equivalent to the original representation.

The third class contains all possible unitary representations. In a finite number of dimensions, these can be decomposed first into factorial representations, and these, in turn, in irreducible ones. Von Neumann[10] has shown that the first step still is possible in infinite dimensions. We can assume, therefore, from the outset that we are dealing with factorial representations.

In the theory of representations of finite dimensions, it is sufficient to determine only the irreducible ones, all others are equivalent to sums of these. Here, it will be necessary to determine all factorial representations. Having done that, we shall know from the above theorem of von Neumann, that all representations are equivalent to finite or infinite sums of factorial representations.

It will be one of the results of the detailed investigation that the inhomogeneous Lorentz group has no "continuous" representations, all representations can be decomposed into irreducible ones. Thus the work of Majorana and Dirac appears to be justified from this point of view a posteriori.

D. *Classification of unitary representations from the point of view of infinitesimal operators*

The existence of an infinitesimal operator of a continuous one parametric (cyclic, abelian) unitary group has been shown by Stone.[11] He proved that the operators of such a group can be written as $\exp(iHt)$ where H is a (bounded or unbounded) hermitean operator and t is the group parameter. However, the Lorentz group has many one parametric subgroups, and the corresponding infinitesimal operators $H_1, H_2, \ldots$ are all unbounded. For every H_i an everywhere dense set of functions ϕ can be found such that $H_i \phi$ can be defined. It is not clear, however, that an everywhere dense set can be found, to all members of which every H can be applied. In fact, it is not clear that one such ϕ can be found.

Indeed, it may be interesting to remark that for an irreducible representation the existence of one function ϕ to which all infinitesimal operators can be applied, entails the existence of an everywhere dense set of such functions. This again has the consequence that one can operate with infinitesimal operators to a large extent in the usual way.

PROOF: Let $Q(t)$ be a one parametric subgroup such that $Q(t)Q(t') = Q(t+t')$. If the infinitesimal operator of all subgroups can be applied to ϕ, the

$$\lim_{t=0} t^{-1}(Q(t)-1)\phi \tag{5}$$

exists. It follows, then, that the infinitesimal operators can be applied to $R\phi$ also where R is an arbitrary operator of the repre-

[11] M. H. Stone, *Proc. Nat. Acad.* **16**, 173, 1930, *Ann. of Math.* **33**, 643, 1932, also J. v. Neumann, *ibid.* **33**, 567, 1932.

sentation: Since $R^{-1}Q(t)R$ is also a one parametric subgroup

$$\lim_{t=0} t^{-}\bigl(R^{-1}Q(t)R-1\bigr)\phi = \lim_{t=0} R^{-1} \cdot t^{-1}\bigl(Q(t)-1\bigr)R\phi$$

also exists and hence also (R is unitary)

$$\lim_{t=0} t^{-1}\bigl(Q(t)-1\bigr)R\phi.$$

Every infinitesimal operator can be applied to $R\phi$ if they all can be applied to ϕ, and the same holds for sums of the kind

$$a_1 R_1 \phi + a_2 R_2 \phi + \ldots + a_n R_n \phi. \tag{6}$$

These form, however, an everywhere dense set of functions if the representation is irreducible.

If the representation is not irreducible, one can consider the set N_0 of such wave functions to which every infinitesimal operator can be applied. This set is clearly linear and, according to the previous paragraph, invariant under the operations of the group (i.e. contains every $R\phi$ if it contains ϕ). The same holds for the closed set N generated by N_0 and also of the set P of functions which are perpendicular to all functions of N. In fact, if ϕ_p is perpendicular to all ϕ_n of N, it is perpendicular also to all $R^{-1}\phi_n$ and, for the unitary character of R, the $R\phi_p$ is perpendicular to all ϕ_n, i.e. is also contained in the set P.

We can decompose thus, by a unitary transformation, every unitary representation into a "normal" and a "pathological" part. For the former, there is an everywhere dense set of functions, to which all infinitesimal operators can be applied. There is no single wave functions to which all infinitesimal operators of a "pathological" representation could be applied.

According to Murray and von Neumann, if the original representation was factorial, all representations into which it can be decomposed will be factorial also. Thus every representation is equivalent to a sum of factorial representations, part of which is "normal," the other part "pathological."

It will turn out again that the inhomogeneous Lorentz group has no pathological representations. Thus this assumption of Majorana and Dirac also will be justified a posteriori. Every unitary representation of the inhomogeneous Lorentz group can be decomposed into normal irreducible representations. It should be stated, however, that the representations in which the unit operator corresponds to every translation have not been determined to date (cf. also Section 3, end). Hence, the above statements are not proved for these representations, which are, however, more truly representations of the homogeneous Lorentz group, than of the inhomogeneous group.

While all these points may be of interest to the mathematician only, the new representation of the Lorentz group which will be described in Section 7 may interest the physicist also. It describes a particle with a continuous spin.

Acknowledgement. The subject of this paper was suggested to me as early as 1928 by P. A. M. Dirac who realized even at that date the connection of representations with quantum mechanical equations. I am greatly indebted to him also for many fruitful conversations about this subject, especially during the years 1934/35, the outgrowth of which the present paper is.

I am indebted also to J. v. Neumann for his help and friendly advice.

3. Summary of Ensuing Sections

Section 4 will be devoted to the definition of the inhomogeneous Lorentz group and the theory of characteristic values and characteristic vectors of a homogeneous (ordinary) Lorentz transformation. The discussion will follow very closely the corresponding, well-known theory of the group of motions in ordinary space and the theory of characteristic values of orthogonal transformations.[12]

---

[12] Cf., e.g., E. Wigner, loc. cit., Chapter III. O. Veblen and J. W. Young, *Projective Geometry*, Boston, 1917. Vol. 2, especially Chapter VII.

It will contain only a straightforward generalization of the method usually applied in those discussions.

In Section 5,* it will be proved that one can determine the physically meaningless constants in the $D(L)$ in such a way that instead of (3a) the more special equation

$$D(L_1)\,D(L_2) = \pm\, D(L_1 L_2) \tag{7}$$

will be valid. This means that instead of a representation up to a factor, we can consider representations up to the sign. For the case that either L_1 or L_2 is a pure translation, Dirac[13] has given a proof of (7) using infinitesimal operators. A consideration very similar to his can be carried out, however, also using only finite transformations.

For representations with a finite number of dimensions (corresponding to an only finite number of linearly independent states), (7) could be proved also if both L_1 and L_2 are homogeneous Lorentz transformations, by a straightforward application of the method of Weyl and Schreier.[14] However, the Lorentz group has no finite dimensional representation (apart from the trivial one in which the unit operation corresponds to every L). Thus the method of Weyl and Schreier cannot be applied. Its first step is to normalize the indeterminate constants in every matrix $D(L)$ in such a way that the determinant of $D(L)$ becomes 1. No determinant can be defined for general unitary operators.

The method to be employed here will be to decompose every L into a product of two involutions $L = MN$ with $M^2 = N^2 = 1$. Then $D(M)$ and $D(N)$ will be normalized so that their squares become unity and $D(L) = D(M)D(N)$ set. It will be possible, then, to prove (7) without going back to the topology of the group.

Sections 6, 7, and 8 will contain the determination of the representations. The pure translations form an invariant subgroup of

* Section 5 and beyond are omitted from the present abstract.

[13] P. A. M. Dirac, mimeographed notes of lectures delivered at Princeton University, 1934/35, page 5a.

[14] H. Weyl, *Mathem. Zeits.* **23**, 271; **24**, 328, 377, 789, 1925; O. Schreier, *Abhandl. Mathem. Seminar Hamburg*, **4**, 15, 1926; **5**, 233, 1927.

the whole inhomogeneous Lorentz group and Frobenius' method[15] will be applied in Section 6 to build up the representations of the whole group out of representations of the subgroup, by means of a "little group." In Section 6, it will be shown on the basis of an as yet unpublished work of J.v. Neumann that there is a characteristic (invariant) set of "momentum vectors" for every irreducible representation. The irreducible representations of the Lorentz group will be divided into four classes.

The momentum vectors of the

1st class are time-like,
2nd class are null-vectors, but not all their components will be zero,
3rd class vanish (i.e., all their components will be zero),
4th class are space-like.

Only the first two cases will be considered in Section 7, although the last case may be the most interesting from the mathematical point of view. I hope to return to it in another paper. I did not succeed so far in giving a complete discussion of the 3rd class. (All these restrictions appear in the previous treatments also.)

In Section 7, we shall find again all known representations of the inhomogeneous Lorentz group (i.e., all known Lorentz invariant equations) and two new sets.

Sections 5, 6, 7 will deal with the "restricted Lorentz group" only, i.e. Lorentz transformations with determinant 1 which do not reverse the direction of the time axis. In section 8, the representations of the extended Lorentz group will be considered, the transformations of which are not subject to these conditions.

---

[15] G. Frobenius, *Sitz. d. Kön. Preuss. Akad.* p. 501, 1898; I. Schur, *ibid.* p. 164, 1906; F. Seitz, *Ann. of Math.* **37**, 17, 1936.

4. Description of the Inhomogeneous Lorentz Group

$A.\Lambda$

An inhomogeneous Lorentz transformation $L = (a, \Lambda)$ is the product of a translation by a real vector a

$$x'_i = x_i + a_i \qquad (i = 1, 2, 3, 4) \tag{8}$$

and a homogeneous Lorentz transformation Λ with real coefficients

$$x'_i = \sum_{k=1}^{4} \Lambda_{ik} x_k. \tag{9}$$

The translation shall be performed after the homogeneous transformation. The coefficients of the homogeneous transformation satisfy three conditions: (1) They are real and Λ leaves the indefinite quadratic form $-x_1^2 - x_2^2 - x_3^2 + x_4^2$ invariant:

$$\Lambda F \Lambda' = F \tag{10}$$

where the prime denotes the interchange of rows and columns and F is the diagonal matrix with the diagonal elements $-1, -1, -1, +1$. (2) The determinant $|\Lambda_{ik}| = 1$ and (3) $\Lambda_{44} > 0$.

We shall denote the Lorentz-hermitean product of two vectors x and y by

$$\{x, y\} = -x_1^* y_1 - x_2^* y_2 - x_3^* y_3 + x_4^* y_4. \tag{11}$$

(The star denotes the conjugate imaginary.) If $\{x, x\} < 0$ the vector x is called space-like, if $\{x, x\} = 0$, it is a null vector, if $\{x, x\} > 0$, it is called time-like. A real time-like vector lies in the positive light cone if $x_4 > 0$; it lies in the negative light cone if $x_4 < 0$. Two vectors x and y are called orthogonal if $\{x, y\} = 0$.

On account of its linear character a homogeneous Lorentz transformation is completely defined if Λv is given for four linearly independent vectors $v^{(1)}, v^{(2)}, v^{(3)}, v^{(4)}$.

From (11) and (10) it follows that $\{v, w\} = \{\Lambda v, \Lambda w\}$ for every pair of vectors v, w. This will be satisfied for every pair if it is satisfied for all pairs $v^{(i)}$, $v^{(k)}$ of four linearly independent vectors. The reality condition is satisfied if $(\Lambda v^{(i)})^* = \Lambda(v^{(i)*})$ holds for four such vectors.

The scalar product of two vectors x and y is positive if both lie in the positive light cone or both lie in the negative light cone. It is negative if one lies in the positive, the other in the negative light cone. Since both x and y are time-like $|x_4|^2 > |x_1|^2 + |x_2|^2 + |x_3|^2$; $|y_4|^2 > |y_1|^2 + |y_2|^2 + |y_3|^2$. Hence, by Schwarz's inequality $|x_4^* y_4| > |x_1^* y_1 + x_2^* y_2 + x_3^* y_3|$ and the sign of the scalar product of two real time-like vectors is determined by the product of their time components.

A time-like vector is transformed by a Lorentz transformation into a time-like vector. Furthermore, on account of the condition $\Lambda_{44} > 0$, the vector $v^{(0)}$ with the components 0, 0, 0, 1 remains in the positive light cone, since the fourth component of $\Lambda v^{(0)}$ is Λ_{44}. If $v^{(1)}$ is another vector[16] in the positive light cone $\{v^{(1)}, v^{(0)}\} > 0$ and hence also $\{\Lambda v^{(1)}, \Lambda v^{(0)}\} > 0$ and $\Lambda v^{(1)}$ is in the positive light cone also. The third condition for a Lorentz transformation can be formulated also as the requirement that every vector in (or on) the positive light cone shall remain in (or, respectively, on) the positive light cone.

This formulation of the third conditions hows that the third condition holds for the product of two homogeneous Lorentz transformations if it holds for both factors. The same is evident for the first two conditions.

From $\Lambda F \Lambda' = F$ one obtains by multiplying with Λ^{-1} from the left and $\Lambda'^{-1} = (\Lambda^{-1})'$ from the right $F = \Lambda^{-1} F (\Lambda^{-1})'$ so that the reciprocal of a homogeneous Lorentz transformation is again such a transformation. The homogeneous Lorentz transformations form a group, therefore.

---

[16] Wherever a confusion between vectors and vector components appears to be possible, upper indices will be used for distinguishing different vectors and lower indices for denoting the components of a vector.

One easily calculates that the product of two inhomogeneous Lorentz transformations (b, M) and (c, N) is again an inhomogeneous Lorentz transformation (a, Λ)

$$(b, M)(c, N) = (a, \Lambda) \tag{12}$$

where

$$\Lambda_{ik} = \sum_j M_{ij} N_{jk}; \qquad a_i = b_i + \sum_j M_{ij} c_j, \tag{12a}$$

or, somewhat shorter

$$\Lambda = MN; \qquad a = b + Mc. \tag{12b}$$

B. Theory of Characteristic Values and Characteristic Vectors of a Homogeneous Lorentz Transformation

Linear homogeneous transformations are most simply described by their characteristic values and vectors. Before doing this for the homogeneous Lorentz group, however, we shall need two rules about orthogonal vectors.

[1] *If* $\{v, w\} = 0$ *and* $\{v, v\} > 0$, *then* $\{w, w\} < 0$; *if* $\{v, w\} = 0$, $\{v, v\} = 0$, *then* w *is either space-like, or parallel to* v (*either* $\{w, w\} < 0$, *or* $w = cv$).

PROOF:

$$v_4^* w_4 = v_1^* w_1 + v_2^* w_2 + v_3^* w_3. \tag{13}$$

By Schwarz's inequality, then

$$|v_4|^2 |w_4|^2 \leqq (|v_1|^2 + |v_2|^2 + |v_3|^2)(|w_1|^2 + |w_2|^2 + |w_3|^2). \tag{14}$$

For $|v_4|^2 > |v_1|^2 + |v_2|^2 + |v_3|^2$ it follows that $|w_4|^2 < |w_1|^2 + |w_2|^2 + |w_3|^2$. If $|v_4|^2 = |v_1|^2 + |v_2|^2 + |v_3|^2$ the second inequality still follows if the inequality sign holds in (14). The equality sign can hold only, however, if the first three components of the vectors v and w are proportional. Then, on account of (13) and both being null vectors the fourth components are in the same ratio also.

[2] *If four vectors* $v^{(1)}, v^{(2)}, v^{(3)}, v^{(4)}$ *are mutually orthogonal and linearly independent, one of them is time-like, three are space-like.*

PROOF: It follows from the previous paragraph that only one of four mutually orthogonal, linearly independent vectors can be time-like or a null vector. It remains to be shown therefore only that one of them *is* time-like. Since they are linearly independent, it is possible to express by them any time-like vector

$$v^{(t)} = \sum_{k=1}^{4} \alpha_k v^{(k)}.$$

The scalar product of the left side of this equation with itself is positive and therefore

$$\left\{\sum_k \alpha_k v^{(k)}, \quad \sum_k \alpha_k v^{(k)}\right\} > 0$$

or

$$\sum_k |\alpha_k|^2 \{v^{(k)}, v^{(k)}\} > 0 \tag{15}$$

and one $\{v^{(k)}, v^{(k)}\}$ must be positive. Four mutually orthogonal vectors are not necessarily linearly independent, because a null vector is perpendicular to itself. The linear independence follows, however, if none of the four is a null vector.

We go over now to the characteristic values λ of Λ. These make the determinant $|\Lambda - \lambda 1|$ of the matrix $\Lambda - \lambda 1$ vanish.

[3] *If λ is a characteristic value, λ^*, λ^{-1} and λ^{*-1} are characteristic values also.*

PROOF: For λ^* this follows the fact that Λ is real. Furthermore, from $|\Lambda - \lambda 1| = 0$ also $|\Lambda' - \lambda 1| = 0$ follows, and this multiplied by the determinants of ΛF and F^{-1} gives

$$|\Lambda F|.|\Lambda' - \lambda 1|.|F|^{-1} = |\Lambda F \Lambda' F^{-1} - \lambda \Lambda| = |1 - \lambda \Lambda| = 0,$$

so that λ^{-1} is a characteristic value also.

[4] *The characteristic vectors v_1 and v_2 belonging to two characteristic values λ_1 and λ_2 are orthogonal if $\lambda_1^* \lambda_2 \neq 1$.*

PROOF:

$$\{v_1, v_2\} = \{Av_1, Av_2\} = \{\lambda_1 v_1, \lambda_2 v_2\} = \lambda_1^* \lambda_2 \{v_1, v_2\}.$$

Thus if $\{v_1, v_2\} \neq 0$, $\lambda_1^* \lambda_2 = 1$.

[5] *If the modulus of a characteristic value λ is $|\lambda| \neq 1$, the corresponding characteristic vector v is a null vector and λ itself real and positive.*

From $\{v, v\} = \{Av, Av\} = |\lambda|^2 \{v, v\}$ the $\{v, v\} = 0$ follows immediately for $|\lambda| \neq 1$. If λ were complex, λ^* would be a characteristic value also. The characteristic vectors of λ and λ^* would be two different null vectors and, because of [4], orthogonal to each other. This is impossible on account of [1]. Thus λ is real and v a real null vector. Then, on account of the third condition for a homogeneous Lorentz transformation, λ must be positive.

[6] *The characteristic value λ of a characteristic vector v of length null is real and positive.*

If λ were not real, λ^* would be a characteristic value also. The corresponding characteristic vector v^* would be different from v, a null vector also, and perpendicular to v on account of [4]. This is impossible because of [1].

[7] *The characteristic v of a complex characteristic value λ (the modulus of which is 1 on account of [5]) is space-like: $\{v, v\} < 0$.*

PROOF: λ^* is a characteristic value also, the corresponding characteristic vector is v^*. Since $(\lambda^*)^* \lambda = \lambda^2 \neq 1$, $\{v^*, v\} = 0$. Since they are different, at least one is space-like. On account of $\{v, v\} = \{v^*, v^*\}$ both are space-like. If all four characteristic values were complex and the corresponding characteristic vectors linearly independent (which is true except if A has elementary divisors) we should have four space-like, mutually orthogonal vectors. This is impossible, on account of [2]. Hence

[8] *There is not more than one pair of conjugate complex characteristic values, if A has no elementary divisors. Similarly, under the same condition, there is not more than one pair λ, λ^{-1} of characteris-*

tic values whose modulus is different from 1. Otherwise their characteristic vectors would be orthogonal, which they cannot be, being null vectors.

For homogeneous Lorentz transformations which do not have elementary divisors, the following possibilities remain:

(a) There is a pair of complex characteristic values, their modulus is 1, on account of [5]

$$\lambda_1 = \lambda_2^* = \lambda_2^{-1}; \qquad |\lambda_1| = |\lambda_2| = 1, \qquad (16)$$

and also a pair of characteristic values λ_3, λ_4, the modulus of which is not 1. These must be real and positive:

$$\lambda_4 = \lambda_3^{-1}; \qquad \lambda_3 = \lambda_3^* > 0. \qquad (16a)$$

The characteristic vectors of the conjugate complex characteristic values are conjugate complex, perpendicular to each other and space-like so that they can be normalized to -1

$$v_1 = v_2^*; \qquad \{v_1, v_2\} = \{v_1 v_1^*,\} = 0$$
$$\{v_1, v_1\} = \{v_2, v_2\} = -1 \qquad (17)$$

those of the real characteristic values are real null vectors, their scalar product can be normalized to 1

$$v_3 = v_3^* \quad v_4 = v_4^* \quad \{v_3, v_4\} = 1$$
$$\{v_3, v_3\} = \{v_4, v_4\} = 0. \qquad (17a)$$

Finally, the former pair of characteristic vectors is perpendicular to the latter kind

$$\{v_1, v_3\} = \{v_1, v_4\} = \{v_2, v_3\} = \{v_2, v_4\} = 0. \qquad (17b)$$

It will turn out that all the other cases in which Λ has no elementary divisor are special cases of (a).

(b) There is a pair of complex characteristic values λ_1, $\lambda_2 = \lambda_1^{-1} = \lambda_1^*$, $\lambda_1 \neq \lambda_1^*$, $|\lambda_1| = |\lambda_2| = 1$. No pair with $|\lambda_3| \neq 1$, however. Then on account of [8], still $\lambda_3 = \lambda_3^*$ which gives with

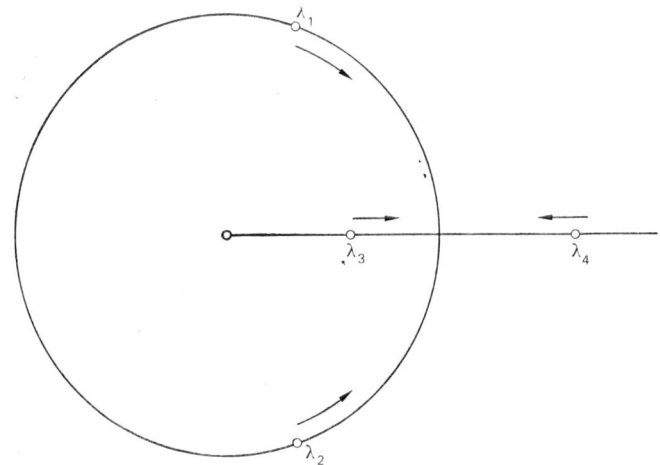

FIG. 1. Position of the characteristic values for the general case (a) in the complex plane. In case (b) λ_2 and λ_4 coincide and are equal 1; in case (c) λ_1 and λ_2 coincide and are either $+1$ or -1. In case (d) both pairs $\lambda_3 = \lambda_4 = 1$ and $\lambda_1 = \lambda_2 = \pm 1$ coincide.

$|\lambda_3| = 1$, $\lambda_3 = \pm 1$. Since the product $\lambda_1 \lambda_2 \lambda_3 \lambda_4 = 1$, on account of the second condition for homogeneous Lorentz transformations, also $\lambda_4 = \lambda_3 = \pm 1$. The double characteristic value ± 1 has two linearly independent characteristic vectors v_3 and v_4 which can be assumed to be perpendicular to each other, $\{v_3, v_4\} = 0$. According to [2], one of the four characteristic vectors must be time-like and since those of λ_1 and λ_2 are space-like, the time-like one must belong to ± 1. This must be positive, therefore $\lambda_3 = \lambda_4 = 1$. Out of the time-like and space-like vectors $\{v_3, v_3\} = -1$ and $\{v_4, v_4\} = 1$, one can build two null vectors $v_4 + v_3$ and $v_4 - v_3$. Doing this, case (b) becomes the special case of (a) in which the real positive characteristic values become equal: $\lambda_3 = \lambda_4^{-1} = 1$.

(c) All characteristic values are real; there is however one pair $\lambda_3 = \lambda_3^*$, $\lambda_4 = \lambda_3^{-1}$, the modulus of which is not unity. Then $\{v_3, v_3\} = \{v_4, v_4\} = 0$ and $\lambda_3 > 0$ and one can conclude for λ_1 and λ_2, as before for λ_3 and λ_4 that $\lambda_1 = \lambda_2 = \pm 1$. This again is a special case of (a); here the two characteristic values of modulus 1 become equal.

(d) All characteristic values are real and of modulus 1. If all of them are $+1$, we have the unit matrix which clearly can be considered as a special case of (a). The other case is $\lambda_1 = \lambda_2 = -1$, $\lambda_3 = \lambda_4 = +1$. The characteristic vectors of λ_1 and λ_2 must be space-like, on account of the third condition for a homogeneous Lorentz transformation; they can be assumed to be orthogonal and normalized to -1. This is then a special case of (b) and hence of (a) also. The cases (a), (b), (c), (d) are illustrated in Fig. 1.

The cases remain to be considered in which Λ has an elementary divisor. We set therefore

$$\Lambda_e v_e = \lambda_e v_e; \qquad \Lambda_e w_e = \lambda_e w_e + v_e. \tag{18}$$

It follows from [5] that either $|\lambda_e| = 1$, or $\{v_e, v_e\} = 0$. We have $\{v_e, w_e\} = \{\Lambda_e v_e, \Lambda_e w_e\} = |\lambda_e|^2 \{v_e, w_e\} + \{v_e, v_e\}$. From this equation

$$\{v_e, v_e\} = 0 \tag{19}$$

follows for $|\lambda_e| = 1$, so that (19) holds in any case. It follows then from [6] that λ_e is real, positive and v_e, w_e can be assumed to be real also. The last equation now becomes $\{v_e, w_e\} = \lambda_e^2 \{v_e, w_e\}$ so that either $\lambda_e = 1$ or $\{v_e, w_e\} = 0$. Finally, we have

$$\{w_e, w_e\} = \{\Lambda_e w_e, \Lambda_e w_e\} = \lambda_e^2 \{w_e, w_e\} + 2\lambda_e \{w_e, v_e\} + \{v_e, v_e\}.$$

This equation now shows that

$$\{w_e, v_e\} = 0 \tag{19a}$$

even if $\lambda_e = 1$. From (19), (19a) it follows that w_e is space-like and can be normalized to

$$\{w_e, w_e\} = -1. \tag{19b}$$

Inserting (19a) into the preceding equation we finally obtain

$$\lambda_e = 1. \tag{19c}$$

[9] *If Λ_e has an elementary divisor, all its characteristic roots are* 1.

From (19c) we see that the root of the elementary divisor is 1 and this is at least a double root. If Λ had a pair of characteristic values $\lambda_1 \neq 1$, $\lambda_2 = \lambda_1^{-1}$, the corresponding characteristic vectors v_1 and v_2 would be orthogonal to v_e and therefore space-like. On account of [5], then $|\lambda_1| = |\lambda_2| = 1$ and $\{v_1, v_2\} = 0$. Furthermore, from $\{w_e, v_1\} = \{\Lambda_e w_e, \Lambda_e v_1\} = \lambda_1 \{w_e, v_1\} + \lambda_1 \{v_e, v_1\}$ and from $\{v_e, v_1\} = 0$ also $\{w_e, v_1\} = 0$ follows. Thus all the four vectors v_1, v_2, v_e, w_e would be mutually orthogonal. This is excluded by [2] and (19).

Two cases are conceivable now. Either the fourfold characteristic root has only one characteristic vector, or there is in addition to v_e (at least) another characteristic vector v_1. In the former case four linearly independent vectors v_e, w_e, z_e, x_e could be found such that

$$\Lambda_e v_e = v_e \qquad \Lambda_e w_e = w_e + v_e$$
$$\Lambda_e z_e = z_e + w_e \qquad \Lambda_e x_e = x_e + z_e.$$

However $\{v_e, x_e\} = \{\Lambda_e v_e, \Lambda_e x_e\} = \{v_e, x_e\} + \{v_e, z_e\}$ from which $\{v_e, z_e\} = 0$ follows. On the other hand

$$\{w_e, z_e\} = \{\Lambda_e w_e, \Lambda_e z_e\} = \{w_e, z_e\} + \{w_e, w_e\} + \{v_e, z_e\} + \{v_e, w_e\}.$$

This gives with (19a) and (19b) $\{v_e, z_e\} = 1$ so that this case must be excluded.

(e) There is thus a vector v_1 so that in addition to (18)

$$\Lambda_e v_1 = v_1 \tag{18a}$$

holds. From $\{w_e, v_1\} = \{\Lambda_e w_e, \Lambda_e v_1\} = \{w_e, v_1\} + \{v_e, v_1\}$ follows

$$\{v_e, v_1\} = 0. \tag{19d}$$

The equations (18), (18a) will remain unchanged if we add to w_e and v_1 a multiple of v_e. We can achieve in this way that the fourth components of both w_e and v_1 vanish. Furthermore, v_1 can be normalized to -1 and added to w_e also with an arbitrary coefficient, to make it orthogonal to v_1. Hence, we can assume that

$$v_{14} = w_{e4} = 0; \quad \{v_1, v_1\} = -1; \quad \{w_e, v_1\} = 0. \quad (19e)$$

We can finally define the null vector z_e to be orthogonal to w_e and v_1 and have a scalar product 1 with v_e

$$\{z_e, z_e\} = \{z_e, w_e\} = \{z_e, v_1\} = 0; \quad \{z_e, v_e\} = 1. \quad (19f)$$

Then the null vectors v_e and z_e represent the momenta of two light beams in opposite directions. If we set $\varLambda_e z_e = av_e + bw_e + cz_e + dv_1$ the conditions $\{z_e, v\} = \{\varLambda_e z_e, \varLambda_e v\}$ give, if we set for v the vectors v_e, w_e, z_e, v_1 the conditions $c = 1$; $b = c$; $2ac - b^2 - d^2 = 0$; $d = 0$. Hence

$$\begin{aligned}\varLambda_e v_e &= v_e & \varLambda_e w_e &= w_e + v_e \\ \varLambda_e v_1 &= v_1 & \varLambda_e z_e &= z_e + w_e + \tfrac{1}{2} v_e.\end{aligned} \quad (20)$$

A Lorentz transformation with an elementary divisor can be best characterized by the null vector v_e which is invariant under it and the space part of which forms with the two other vectors w_e and v_1 three mutually orthogonal vectors in ordinary space. The two vectors w_e and v_1 are normalized, v_1 is invariant under $\varLambda_e$ while the vector v_e is added to w_e upon application of $\varLambda_e$. The results of the application of $\varLambda_e$ to a vector which is linearly independent of v_e, w_e and v_1 is, as we saw, already determined by the expressions for $\varLambda_e v_e$, $\varLambda_e w_e$ and $\varLambda_e v_1$.

The $\varLambda_e(\gamma)$ which have the invariant null vector v_e and also w_e (and hence also v_1) in common and differ only by adding to w_e different multiples γv_e of v_e, form a cyclic group with $\gamma = 0$, the unit transformation as unity:

$$\varLambda_e(\gamma)\varLambda_e(\gamma') = \varLambda_e(\gamma + \gamma').$$

The Lorentz transformation $M(\alpha)$ which leaves v_1 and w_e invariant but replaces v_e by αv_e (and z_e by $\alpha^{-1} z_e$) has the property of transforming $\Lambda_e(\gamma)$ into

$$M(\alpha)\, \Lambda_e(\gamma)\, M(\alpha)^{-1} = \Lambda_e(\alpha\gamma). \qquad (+)$$

An example of $\Lambda_e(\gamma)$ and $M(\alpha)$ is

$$\Lambda_e(\gamma) = \begin{Vmatrix} 1 & 0 & 0 & 0 \\ 0 & 1 & \gamma & \gamma \\ 0 & -\gamma & 1-\tfrac{1}{2}\gamma^2 & -\tfrac{1}{2}\gamma^2 \\ 0 & \gamma & \tfrac{1}{2}\gamma^2 & 1+\tfrac{1}{2}\gamma^2 \end{Vmatrix};$$

$$M(\alpha) = \begin{Vmatrix} 1 & 0 & 0 & 0 \\ 0 & 1 & 0 & 0 \\ 0 & 0 & \tfrac{1}{2}(\alpha+\alpha^{-1}) & \tfrac{1}{2}(\alpha-\alpha^{-1}) \\ 0 & 0 & \tfrac{1}{2}(\alpha-\alpha^{-1}) & \tfrac{1}{2}(\alpha+\alpha^{-1}) \end{Vmatrix}.$$

These Lorentz transformations play an important rôle in the representations with space like momentum vectors.

A behavior like $(+)$ is impossible for finite unitary matrices because the characteristic values of $M(\alpha)^{-1}\Lambda_e(\gamma)M(\alpha)$ and $\Lambda_e(\gamma)$ are the same—those of $\Lambda_e(\gamma\alpha) = \Lambda_e(\gamma)^\alpha$ the αth powers of those of $\Lambda_e(\gamma)$. This shows very simply that the Lorentz group has no true unitary representation in a finite number of dimensions.

C. *Decomposition of a Homogeneous Lorentz Transformation into Rotations and an Acceleration in a Given Direction*

The homogeneous Lorentz transformation is, from the point of view of the physicist, a transformation to a uniformly moving coordinate system, the origin of which coincided at $t = 0$ with the origin of the first coordinate system. One can, therefore, first perform a rotation which brings the direction of motion of the second system into a given direction—say the direction of the third axis—and impart it a velocity in this direction, which will bring it to rest. After this, the two coordinate systems can differ only in a rotation. This means that every homogeneous Lorentz transformation can be decomposed in the following way[17]

$$\Lambda = RZS \tag{21}$$

where R and S are pure rotations, (i.e. $R_{i4} = R_{4i} = S_{i4} = S_{4i} = 0$ for $i \neq 4$ and $R_{44} = S_{44} = 1$, also $R' = R^{-1}$, $S' = S^{-1}$) and Z is an acceleration in the direction of the third axis, i.e.

$$Z = \begin{Vmatrix} 1 & 0 & 0 & 0 \\ 0 & 1 & 0 & 0 \\ 0 & 0 & a & b \\ 0 & 0 & b & a \end{Vmatrix}$$

with $a^2 - b^2 = 1$, $a > b > 0$. The decomposition (21) is clearly not unique. It will be shown, however, that Z is uniquely determined, i.e. the same in every decomposition of the form (21).

In order to prove this mathematically, we choose R so that in $R^{-1}\Lambda = I$ the first two components in the fourth column $I_{14} = I_{24} = 0$ become zero: R^{-1} shall bring the vector with the components Λ_{14}, Λ_{24}, Λ_{34} into the third axis. Then we take $I_{34} = (\Lambda_{14}^2 + \Lambda_{24}^2 + \Lambda_{34}^2)^{1/2}$ and $I_{44} = \Lambda_{44}$ for b and a to form Z; they satisfy the equation $I_{44}^2 - I_{34}^2 = 1$. Hence, the first three components of the fourth column of $J = Z^{-1}I = Z^{-1}R^{-1}\Lambda$ will become zero and $J_{44} = 1$, because of $J_{44}^2 - J_{14}^2 - J_{24}^2 - J_{34}^2 = 1$. Furthermore, the first three components of the fourth row of J will vanish also, on account of $J_{44}^2 - J_{41}^2 - J_{42}^2 - J_{43}^2 = 1$, i.e. $J = S = Z^{-1}R^{-1}\Lambda$ is a pure rotation. This proves the possibility of the decomposition (21).

The trace of $\Lambda\Lambda' = RZ^2R^{-1}$ is equal to the trace of Z^2, i.e. equal to $2a^2 + 2b^2 + 2 = 4a^2 = 4b^2 + 4$ which shows that the a and b of Z are uniquely determined. In particular $a = 1$, $b = 0$ and Z the unit matrix if $\Lambda\Lambda' = 1$, i.e. Λ a pure rotation.

It is easy to show now that the group space of the homogeneous Lorentz transformations is only doubly connected. If a continuous series $\Lambda(t)$ of homogeneous Lorentz transformations is given, which is unity both for $t = 0$ and for $t = 1$, we can decompose it

[17] Cf., e.g., L. Silberstein, *The Theory of Relativity*, London, 1924, p. 142.

according to (21)
$$\Lambda(t) = R(t) Z(t) S(t). \tag{21a}$$

It is also clear from the foregoing, that $R(t)$ can be assumed to be continuous in t, except for values of t, for which $\Lambda_{14} = \Lambda_{24} = \Lambda_{34} = 0$, i.e. for which Λ is a pure rotation. Similarly, $Z(t)$ will be continuous in t and this will hold even where $\Lambda(t)$ is a pure rotation. Finally, $S = Z^{-1}R^{-1}\Lambda$ will be continuous also, except where $\Lambda(t)$ is a pure rotation.

Let us consider now the series of Lorentz transformations
$$\Lambda_s(t) = R(t) Z(t)^s S(t) \tag{21b}$$
where the b of $Z(t)^s$ is s times the b of $Z(t)$. By decreasing s from 1 to 0 we continuously deform the set $\Lambda_1(t) = \Lambda(t)$ of Lorentz transformations into a set of rotations $\Lambda_0(t) = R(t)S(t)$. Both the beginning $\Lambda_0(0) = 1$ and the end $\Lambda_s(1) = 1$ of the set remain the unit matrix and the sets $\Lambda_s(t)$ remain continuous in t for all values of s. This last fact is evident for such t for which $\Lambda(t)$ is not a rotation: for such t all factors of (21b) are continuous. But it is true also for t_0 for which $\Lambda(t_0)$ is a rotation, and for which, hence $Z(t_0) = 1$ and $\Lambda_s(t_0) = \Lambda_1(t_0) = \Lambda(t_0)$. As $Z(t)$ is everywhere continuous, there will be a neighborhood of t_0 in which $Z(t)$ and hence $Z(t)^s$ is arbitrarily close to the unit matrix. In this neighborhood $\Lambda_s(t) = \Lambda(t). S(t)^{-1}Z(t)^{-1}Z(t)^s S(t)$ is arbitrarily close to $\Lambda(t)$; and, if the neighborhod is small enough, this is arbitrarily close to $\Lambda(t_0) = \Lambda_s(t_0)$.

Thus (21b) replaces the continuous set $\Lambda(t)$ of Lorentz transformations by a continuous set of rotations. Since these form an only doubly connected manifold, the manifold of Lorentz transformations can not be more than doubly connected. The existence of a two valued representation[18] shows that it is actually doubly and not simply connected.

[18] Cf. H. Weyl, *Gruppentheorie und Quantenmechanik*, 1st ed. Leipzig, 1928, pages 110–114, 2nd ed. Leipzig, 1931, pages 130–133. It may be interesting to remark that essentially the same isomorphism has been recognized already by L. Silberstein, *loc. cit.*, pages 148–157.

We can form a new group[14] from the Lorentz group, the elements of which are the elements of the Lorentz group, together with a way $\Lambda(t)$, connecting $\Lambda(1) = \Lambda$ with the unity $\Lambda(0) = E$. However, two ways which can be continuously deformed into each other are not considered different. The product of the element "Λ with the way $\Lambda(t)$" with the element "I with the way $I(t)$" is the element ΛI with the way which goes from E along $\Lambda(t)$ to Λ and hence along $\Lambda I(t)$ to ΛI. Clearly, the Lorentz group is isomorphic with this group and two elements (corresponding to the two essentially different ways to Λ) of this group correspond to one element of the Lorentz group. It is well known[18] that this group is holomorphic with the group of unimodular complex two dimensional transformations.

Every continuous representation of the Lorentz group "up to the sign" is a singlevalued, continuous representation of this group. The transformation which corresponds to "Λ with the way $\Lambda(t)$" is that $d(\Lambda)$ which is obtained by going over from $d(E) = d(\Lambda(0)) = 1$ continuously along $d(\Lambda(t))$ to $d(\Lambda(1)) = d(\Lambda)$.

D. The Homogeneous Lorentz Group is Simple

It will be shown, first, that an invariant subgroup of the homogeneous Lorentz group contains a rotation (i.e. a transformation which leaves x_4 invariant). We can write an arbitrary alement of the invariant subgroup in the form RZS of (21). From its presence in the invariant subgroup follows that of $S \cdot RZS \cdot S^{-1} = SRZ = TZ$. If X_π is the rotation by π about the first axis, $X_\pi Z X_\pi = Z^{-1}$ and $X_\pi TZ X_\pi^{-1} = X_\pi T X_\pi X_\pi Z X_\pi = X_\pi T X_\pi Z^{-1}$ is contained in the invariant subgroup also and thus the transform of this with Z, i.e. $Z^{-1} X_\pi T X_\pi$ also. The product of this with TZ is $T X_\pi T X_\pi$ which leaves x_4 invariant. If $T X_\pi T X_\pi = 1$ we can take $T Y_\pi T Y_\pi$. If this is the unity also $T X_\pi T X_\pi = T Y_\pi T Y_\pi$ and T commutes with $X_\pi Y_\pi$, i.e is a rotation about the third axis. In this case the space-like (complex) characteristic vectors of TZ in the plane of the first two coordinate axes. Transforming TZ by an acceleration in the direction of the first coordinate axis we obtain a new element of the invariant subgroup for which the space-like characteristic vector will have a not vanishing fourth component. Taking this for RZS we can transform it with S again to obtain a new $SRZ = TZ$. However, since S leaves x_4 invariant, the fourth component of the space like characteristic vectors of this TZ will not vanish and we can obtain from it by the procedure just described a rotation which must be contained in the invariant subgroup.

It remains to be shown that an invariant subgroup which contains a rotation, contains the whole homogeneous Lorentz group. Since the three-dimensional rotation group is simple, all rotations must be contained in the invariant subgroup. Thus the rotation by π around the first axis X_π and also its transform with Z and also

$$ZX_\pi Z^{-1} . X_\pi = Z . X_\pi Z^{-1} X_\pi = Z^2$$

is contained in the invariant subgroup. However, the general acceleration in the direction of the third axis can be written in this form. As all rotations are contained in the invariant subgroup also, (21) shows that this holds for all elements of the homogeneous Lorentz group.

If follows from this that the homogeneous Lorentz group has apart from the representation with unit matrices only true representations. If follows then from the remark at the end of part B, that these have all infinite dimensions. This holds even for the two-valued representations to which we shall be led in Section 5 equ. (52D), as the group elements to which the positive or negative unit matrix corresponds must form an invariant subgroup also, and because the argument at the end of part B holds for two-valued representations also. One easily sees furthermore from the equations (52B), (52C) that it holds for the inhomogeneous Lorentz group equally well.

INDEX

Aberration 5, 34, 207
Abraham 133, 141, 147, 219
Acceleration 42, 137, 215
Addition of velocities 230
Alpha-particle 260
Anderson 74, 255
Angular momentum 76, 247
Arago 6
Aristotle 3

Besso 218
Bohr 59, 60
Bondi 88
Born 64, 69
Brace 119, 139
Bradley 4, 34
Bucherer 146

Cauchy 114
Chadwick 258
Composition (of velocities) 201
Contravariant vector 22
Copernicus 148
Covariant vector 22

Darwin 237, 253, 254
Davisson 60
De Broglie 60
Decomposition (of a Lorentz transformation) 289
Descartes 24
Dirac 48, 69, 70, 72, 237, 265
Doppler shift 36
Doppler's principle 207
Double refraction 139
Duplexity phenomena 238, 245

Einstein 14, 34, 187, 230
Einstein's convention 16
"Elastic" forces 137
Electric moment 136
Electromagnetic momentum 132
Electron 237
Elsasser 60
Energy (of light rays) 209
Energy levels 249
Euler 4, 34, 265
Euler–Lagrange equations 26

Feather 258
Fitzgerald 119, 146
Fizeau 6, 18, 37, 229
Force 137, 215
Fresnel 6, 36, 104, 145
Fresnel's coefficient 229

Galileo 3
Geometrical object 21
Gordon 61, 72, 238
Gordon–Klein interpretation 239
Goudsmit 237
Gravitation 172
Green 114

Hamiltonian 29, 46, 238
Hamilton–Jacobi equation 63
Harmonic oscillator 65
Heisenberg 64, 69, 72, 265
Helmholtz 113
Hesse 12

Inertial frame 13
Infinitesimal operators 274

Inhomogeneous Lorentz group 265
Invariants 175
Isochronous transformations 20

Jaumann 233
Jordan 69, 239

Kaufmann 44, 141
Kelvin 12, 106
Kepler 149
Kinetic energy 217
Klein 61, 72, 74, 239, 269
Kohlrausch 9
Kunsman 60

Landsberg 39
Langevin 146
Laplace 148
Larmor 12, 105, 219
Laue 233
Least action 156
Least time 24, 61
Lie 170
Longitudinal and transverse masses 141
Longitudinal mass 134, 215
Lorentz 34, 119, 146, 219, 230
Lorentz contraction 146
Lorentz group 20, 169
Lorentz transformation 34, 71, 146

MacCullagh 114
Magnetic dielectric 39
Magnetic insulator 219
Majorana 266
Mass 215
Maxwell 9, 39, 57, 92, 105, 187
Maxwell equations 203
Measured mass 44
Michelson 6, 91, 119, 121, 139, 140, 231
Miller 7

Millikan 256
Milne 16
Minkowski 20
Molecular motion 140
Morley 6, 231
Motion 3
Murray 272

Navier 114
Neumann 265
Newton 4, 148
Newtonian transformation 12
Noble 119, 139
Noether 27, 46

Old quantum theory 59

Pauli 237, 253, 254
Planck 58, 66, 215
Poincaré 34, 121, 145
Poisson 114
Poisson bracket 48, 69
Pressure (of radiation) 209
Proca 269
Proper orthogonal group 20

Quantum theory 57

Rayleigh 58, 119, 139
Reflection 23
Refraction 23
Relativity postulate 145
Riemann 8
Rund 52
Rutherford 59, 60

Scattering 260
Schreier 277
Schrödinger 61, 72, 266
Schrödinger's equation 64
Serber 265

INDEX

Shankland 7
Signal 14
Silberstein 290
Simultaneity 14, 188
Simultaneous events 189
Skobeltzyn 257
Snell's law 24
Spin 77
Spinor 81
Stationary state 65
Stefan's law 57
Stokes 103
Stone 274
Symmetry 30
Synchronism 190

Tensor 22
Thomas factor 254
Thompson, J. J. 58
Townsend 59
Transverse mass 134, 215

Trouton 119, 139
Two-valued representation 80

Uhlenbeck 237
Unitary representations 265, 274

Variational principle 23, 45
Virtual work 160
Von Neumann 272

Weber 9
Weinstein 233
Weyl 277, 291
Whitrow 15
Wien 57
Wigner 81, 265
Wilson 39, 219

Zeeman 37, 229